# Marine and Offshore Engineering

# Marine and Offshore Engineering

*Editor*
***T.V. Ramakrishnan***

2015
**Gene-Tech Books**
**New Delhi - 110 002**

**ISBN 81-89729-10-1**

***Published by*** : **GENE-TECH BOOKS**
**4762-63/23, Ansari Road, Darya Ganj,**
**NEW DELHI - 110 002**
**Phone: 41562849**
**e-mail: genetechbooks@yahoo.co.in**

***Printed at*** : **Chawla Offset Printers**
**New Delhi – 110 052**

**PRINTED IN INDIA**

# *Preface*

Offshore construction, with its many current opportunities and its tremendous demands and changes, is emerging as one of the most exciting fields of engineering practice, one which will test human kinds ability to rise to new heights of skill and courage.

The book is written to serve as a guide and reference for practicing engineers both designers and contractors and also for students and teachers.

Today marine engineering has got much more importance. The range of offshore and marine structures is very great most structures in harbours, on coasts, in estuaries and deep rivers are carried out by public bodies. Marine projects have special problems, requiring exemption of specific requirements. Many offshore projects relate to the development of an ocean resource especially oil and gas.

I shall regard my efforts amply rewarded if the book is received with the enthusiasm.

I wish to express my sincere thanks to the publisher's especially to Mr. Anil Mittal for his continued interest in bringing out this edition.

**Author**

# Contents

# Chapter 1

# Introduction

The oceans present a unique set of environmental conditions, which dominate, the methods, support and procedures to be employed in construction offshore. Let us try to understand the importance of oceans.

The oceans are the dominant features of Earth, comprising more than two thirds of its surface, stabilizing its temperature so that life as we know it can exist, providing the water vapour which later falls as rain on the continental "islands," the original source of life and the ultimate collector or sink of all surficial matter, including waste. Oceans have been both a barrier and a conduit over which people and goods have moved with relative ease, spreading culture while garnering Earth's remote resources.

Yet the ocean is fiercely inhospitable, making us dependent on land bases for support. Storm waves have destroyed even the largest vessels, as well as the puny attempts of humans to protect the coastline from the oceans attack. The northernmost ocean, the Arctic, is almost completely covered with perpetual sea ice, while the southern, the Antarctic, carries with it huge tabular icebergs that stretch beyond the horizon.

Opportunity and challenge, safety and terror, wealth and destruction: these are the paradoxes of the seas.

Since before recorded history, oceans have been used for transport, for food, for conquest, and for waste disposal. The

Phoenicians sailed as far as Norway to the North and Capetown to the South, perhaps even on to South America; the Polynesians crossed the Pacific to where they sighted the great wall of the Andes, which to them marked the "end of the world," and to Japan and Indonesia; and navigators from Kerala reached Africa and Indonesia, completing early man's circumnavigation of the globe. Much later came the Arabian sailors whose sea empire extended from West Africa to the Philippines; the Vikings, who sailed to Venice and Canada; and eventually the Western European navigators of the Age of Exploration, who challenged the utmost corners of the globe, including both the Arctic and the Antarctic. Today, more than 30,000 ships ply the trade routes of the world.

Mahan's brilliant insights in *The Influence of Sea Power on the History of the World* (1890) demonstrated the decisive role that has been played by the navies of nations who strove either to dominate the world or repel the challenger. As Mahan points out, it was the Greek sea victory at Artemis that blunted the expansion of the Persians; the Roman domination of the Mediterranean that forced Hannibal to his audacious but futile march through Spain and across the Alps to try to break Rome's stranglehold on Carthaginian trade; Drake's defeat of the Spanish Armada and Nelson's victory at Trafalgar that eventually led to Britain's worldwide empire; and the temporary repulse of the British fleet by the French navy which enabled Washington to force surrender of the food- and munitions-starved Cornwallis. Similarly, it was the U.S. Navy's destruction of the Japanese fleet that led to victory in the Pacific World War II.

The oceans have been regarded until recently as an inexhaustible source of food, fishermen need only be clever enough to trap the fish which roam along its coastlines and its great internal rivers where the cold water, rich in nutrients, intermixes with the warm. Fishermen have learned to survive the storm waves, hurricane winds, dense fogs, and "black" ice that have destroyed their less-able predecessors.

The great ocean basins, long thought to be relatively simple, with stable slopes and flat seafloors, have turned out to be exceedingly complex and dynamic. The study of plate tectonics has revealed the underlying mechanism by which seafloors spread and sediments are eventually subducted. The seafloor is marked by deep canyons and steep escarpments. Great volcanic mountains rise above the ocean floor far higher than Everest rises above its base. Seamounts which have not yet reached the surface or which have been eroded or submerged below it now sport crowns of coral.

Among these oceans float the continents, whose margins extend well beneath the sea. That which essentially extends the continents out under the sea is known as the continental shelf, an area rich in sediments washed off the continents and eroded from the shore to be deposited here in relatively concentrated zones. At the outer edge of the continents is the continental slope, dipping down to the abyssal plain (whose surface is often far from planar). Sediments accumulated on the shelves periodically flow down the slopes, as turbidity currents, to form giant fans at their base.

The shelves and slopes are then inherently unstable, geologically speaking, at least in their surficial deposits. Many of the most striking geological features have occurred during episodic events. As opposed to gradual continuous erosion and deposition, these episodic events include gigantic submarine landslides and turbidity current flows. Similarly, coasts and rivers are periodically altered drastically by episodic events.

A basic property of the oceans, affecting all human activities thereon, is their vastness, their "illimitable expanse" which necessitates long-distance transport of all materials, structures, equipment, and personnel. There are no easy geographic reference points, no stable support for adjoining activity or storing of supplies. This problem of logistics dominates all considerations of construction activities and integrates construction with the transport functions upon which it so heavily depends. This same concern for logistical support occurs to a lesser degree on all marine projects.

The seas have long presented a curious contrast between freedom and politically imposed controls. As a result, a patchwork set of arrangements has emerged. The recent Law of the Sea conferences were an attempt to establish a more logical and politically viable legal basis for the rapidly expanding development of the oceans.

Freedom of navigation, except in a narrow zone close to the shores, and freedom of innocent passage through straits have long been established and enforced by the world's great sea powers. More recently, freedom of scientific research has been promulgated, only to founder on the narrow distinctions among research, exploration for mineral resources, exploitation of pharmacological compounds, and military intelligence.

While the Law of the Sea Treaty has been signed by several nations, it has been rejected by the United States because of inclusion of the concept of a supranational seabed authority to be granted jurisdiction and concomitant expropriation-like rights over seabed mineral resources and the technology used to produce them. Despite the current lack of treaty ratification, however, most of the treaty's other provisions are coming into being as common law, through voluntary observance and through unilateral proclamations of clauses similar to those of the treaty, such as the establishment of 200-mile-wide economic zones.

The most immediate results pertain to control of fishing in these vastly expanded national jurisdictions and the right to produce offshore oil and gas. For example, the United States has by one stroke increased by almost 25 per cent the area of the globe over which it asserts jurisdiction.

Political jurisdiction in the Arctic remains confused, with some of the nations bordering the Arctic asserting the sector theory, that is, the extension of a meridian from their northernmost land boundary directly to the North Pole. The five nations bordering the Arctic Ocean are Greenland (which belongs to Denmark), Canada, the United States (in Alaska), Norway, and Russia, the sector of the last extending almost halfway around the globe. Russia claims the shallow waters of the Barents, Kara, and East Siberian Seas as territorial seas. Canada similarly claims that the channels between the Arctic islands are territorial waterways, whereas the United States asserts that they are international straits with free right of passage.

The Antarctic seas, south of the 60$^{th}$ parallel, remain an anomaly under a regional authority set up under U.N. provisions. The Antarctic Treaty Organization, originally formed to prohibit military use of Antarctica and to foster exchange of scientific information, has more recently been expanded by the establishment of a "Living Resources" regime, primarily to control the exploitation of krill in the cold, upwelling waters around the continent. Whales and seals are currently protected under a parallel Treaty on Marine Mammals. Similarly, a "Mineral Resources" regime constrains development of potential petroleum resources such as the vast submarine sediments of the Ross, Weddell, and Bellingshausen Seas. With more practicable approach to rights and obligations, the protection of the operator, and the sharing of gain, this could eventually set a pattern for revision of the objectionable provisions of the Law of the Sea Treaty.

Closer to the continental shores, within those areas of national jurisdiction, is a 12-mile zone under full control of the adjoining nation. In the United States this is further subdivided by a 3-mile zone under the jurisdiction of the adjoining state. This latter is administered under the provisions of the Coastal Zone Management Act, which takes into account the onshore impacts of offshore activities as well as the direct activities themselves.

The environmental impact statements, carried out under the laws governing these several zones surrounding the coasts, have resulted in a series of agreements concerning specific projects, including their construction. These constraints may affect the procedures, methods, and sequence of construction. They carry the force of law. Of particular interest to construction are those constraints which relate to dredging and dredge disposal. Discharge of oil in harbors, and disposal or capping of contaminated sediments, is a major concern. On our rivers and estuaries, migrations of fish may restrict the times when operations maybe carried out, whereas in harbors, the prohibited zones and periods may relate to breeding and nesting times.

Interest in the life of the sea has expanded in recent years beyond the fascination of Moby Dick and the locating of desirable schools of fish to a deep concern for all living creatures, especially those of the sea, which share our common source of life and which have evolved along parallel lines to relatively high levels of intelligence. As with all newly emerging concerns, there have perhaps been excesses of zeal, but the underlying recognition that the life of the sea must be protected from wholesale depredation has become a basic ethical tenet of our society.

Thus, construction activities in the ocean, especially those in the coastal zones, must take cognizance of ecological and environmental constraints, whether these be limitations on noise generated in the water column by dredging, which may affect the navigational and communications abilities of marine mammals, or pollution by persistent chemical discharges. Paradoxically, the aesthetically and legally unforgivable presence of a sheen of oil on the water may also be the least environmentally harmful, due to its rapid biodegradation into edible protein. Massive oil spills, however, cause extensive damage to the biota of the coasts, but fortunately the damage is not permanent. Perhaps the worst effects occur in the wetlands which adjoin the estuaries.

The coasts are the boundary between the continents and the oceans. This interface is constantly undergoing change; uplifting as an active plate subducts under the continental margin, eroding under the constant pounding of the surf, or accreting from sediment discharge, while periodically discharging immense flows of sand or mud that scour out great submarine canyons.

The great rivers of the world drain the continents, providing fresh water for human and animal consumption, for agriculture and industry. Their flow, however, is not uniform but characterized by periods of low water offset by raging floods which often devastate the adjacent lands yet leave behind the fertile silts and clays. These rivers also provide the easiest and most economical roadways into the interior of the continents; their navigation has been the mainstream of commerce throughout history.

Rivers empty into harbors, around which man has built his great cities. The harbors are ports of refuge for ships from the storms of the oceans, and they are the locale of trans-shipment from ocean going vessels to land and river transport. As cities have grown, clustered mainly at the junctures where great rivers meet the ocean, so have the problems of waste disposal grown, whether it be sewage, the effluent from industrial processes, the runoff of waste oil from urban lands and of nutrients from farmlands, or the warm water discharges from power plants. The sea has been a compliant receiver, quick to disperse and dilute all but the most toxic wastes. The oils have been consumed by bacteria, and most of the excess minerals precipitated to the seafloor.

The above describes the state of the ocean until recent times, the latter half of the 20$^{th}$ century. Now suddenly humankind has burst out with explosive force, increasing both population and human activities at an exponential rate. In the forefront of this revolutionary expansion, this "big bang" of cultural spreading, has been the technological exploitation of the oceans.

In the field of transport, we see new ship types and modes, from containerization to catamarans, hovercraft, and very-large crude oil carriers (VLCCs), while kilometer-long "trains" of barges ply inland rivers made navigable by locks and dams. In fishing, we see electronic search, sea ranching, and the beginning of exploitation of Antarctic krill, that tiny shrimp whose numbers render it the most abundant source of protein on Earth. Even though we live in the Space Age, it remains the seas in which military might dominates, for the nuclear-powered submarine with its awesome destructive power lurks almost undetectable in the ocean depths or underneath sea ice cover. Waste discharges continue, but now there is a global awareness of the need for at least primary treatment and mechanical dispersal to avoid over-concentrations along the vulnerable coasts.

The thermal attributes of the ocean, as a source of cooling water, a sink for warm water, and even a potential source of energy, have long been recognized. Although ocean thermal energy conversion (OTEC) projects are not currently economically viable, their technological feasibility has been demonstrated. In the long term, it will probably be the unlimited source of cooling water of the oceans combined with their capacity to accept discharges that will lead to seaborne industrial processing plants on a large scale.

It has only been in the latter half of the 20th century that full recognition has been given to the oceans and their sediments as a major source of mineral wealth, both hard minerals and petroleum. Offshore oil and gas now supply almost one third of the world's energy needs: in fact, it has been stated by the U.S. Geological Survey that the offshore sedimentary basins within the U.S. Economic Zone hold forth the greatest potential for major new discoveries.

An immense amount of publicity has been given to the manganese nodules which cover large areas of tropical and subtropical seafloors. More recently, the scientific world has been excited by the discovery of the thermal vents from seafloor rifts, with their strange new forms of life and their apparently rich deposits of polysulfide minerals. Extraction of soluble minerals from the sea has been carried out since prehistoric times: salt (sodium chloride) and in modern times, magnesium and bromine.

Coastal sediments are also rich deposits of precious minerals such as gold, tin, and probably chromium and platinum. Seabed mining of such unsophisticated minerals as sand and gravel is of major importance in Japan, many European countries, and the Arctic. However, because of the tremendous economic importance of offshore oil and gas and the concentrated development of technology for their exploitation, much of the recent marine construction practice has been devoted to the installation of facilities to serve the needs of the petroleum industry.

## Chapter 2

# The Marine Working Environment

### Introduction

The demands of the marine working environment, coupled with the demand for large-scale structures, have led to the development of a great many types of specialized and advanced construction equipment. Indeed, the response of equipment manufacturers and constructors has been rapid and effective. The availability of construction equipment of greater capabilities has in turn played a major role in altering construction methods and in making technically feasible and economically justifiable complex structures in extremely demanding environments. These developments will continue as industrial development, principally the offshore petroleum industry, military requirements, and maritime commerce, continue their current rate of growth.

The major construction equipment has been designed to work in and under the sea and hence has drawn heavily on naval architecture to ensure serviceability and stability as well as limited and predictable motion response under the prevailing marine and offshore conditions. This extension from conventional barges and ships, directed primarily for transport, to construction, drilling, and dredging operations has in turn forced the naval architectural profession to develop a methodology adaptable to a wide variety of configurations and dynamic forces. While transport hurricanes and icebreakers follow open leads in the pack ice, fixed structures must survive the full brunt of such environmental extremes.

Life safety must be paramount in offshore operations. The nature of the work is inherently demanding and dangerous. The equipment must be designed not only for serviceability but also for safe operations.

Marine and especially offshore equipment is very expensive: each hour has a high value in ownership or rental, plus high operating costs. Therefore, the equipment must be designed with reliability and redundancy. As a general rule, it should be capable of efficient operations in 70 per cent or more of the days in the working season. Construction engineers must understand the capabilities and limitations of the equipment they use. They must be alert to detect early signs of problems before they develop to catastrophic proportions. Thus a full understanding of equipment performance is essential. In subsequent subsections of this chapter, principal generic types of marine and offshore construction equipment will be discussed.

The marine construction industry has been subject to dramatic cyclic variations, from over demand to recession. In times such as those in which this chapter is being written (1998), when the demand for large specialized equipment exceeds the supply, two responses have developed. One is the placement of orders with shipyards and crane manufacturers for new construction of the standard offshore equipment, upgraded to allow its use in deeper water and in exposed environments. The other, very interesting development has been that in which existing equipment is being modified and new procedures are being developed in order to perform tasks which hitherto were only possible with large conventional equipment. These latter are making extensive use of the newly developed hydraulic jacks, with long strokes, high capacity, and the ability to accommodate transverse relative motion by means of rollers and low-friction materials such as Teflon. For inshore marine operations, such as bridges and locks and dams, these same two contrary approaches are being employed. This is especially true where physical constructions limit maneuverability and draft limits access.

There are a number of basic considerations applicable to all offshore construction equipment. These are motion response, buoyancy, draft, freeboard, stability, and damage control.

## The Offshore Construction Barge

An offshore construction barge must be long enough to have minimal pitch and surge response to the waves in which it normally works, wide enough in beam to have minimum roll, and deep enough to have adequate bending strength against hog, sag, and torsion, as well as adequate freeboard. The deck plating must be sufficiently continuous to enable it to resist the membrane compression, tension, and torsion introduced by wave loading.

Impact loadings can come from wave slam on the bow, from ice, and from boats and other barges hitting against the sides. Unequal loads may be incurred in bending of the bottom hull plates during intentional or accidental grounding and of the deck plates due to cargo loads. Corrosion may reduce the thickness of hull plates.

The internal structure of a barge is subdivided by longitudinal and transverse bulkheads. Because of the relatively high possibility of rupturing of a side plate, with consequent flooding of the adjacent compartment, the longitudinal bulkheads are usually spaced at the middle third of the beam. A single centerline bulkhead could allow flooding of one entire side, causing excessive heel and possible capsizing.

Longitudinal bulkheads plus the two sides provide the longitudinal shear strength of the barge. The transverse bulkheads are usually spaced with one just aft of the bow (the collision bulkhead), one forward of the stern, and one or more in the midships region. These provide the transverse shear strength. Quartering waves produce torsion as well as bending in both planes. The torsional shear runs around the girth of the vessel: sides, deck, and bottom.

Typical offshore barges run from 80 to 160 m in length. Width should be one-third to one fifth the length. Depth will typically run from 1/12 to 1/15 the length. Such ratios have been found to give a reasonable balanced structural performance under wave loadings. Inland barges, subjected to minimal wave loadings and required for operations in shallow water may have depths as low as 1/20 of the length. They may be stiffened by external trussing. Shallow-depth barges are often used in rivers and lakes; these can be very hazardous in exposed locations where not only high quasi-static bending stresses can develop but also dynamic amplification and resonance.

Offshore barges typically have natural periods in roll of 5 to 7 s. This is unfortunately the typical period of wind waves; hence resonant response does occur. Fortunately, damping is very high, so that while motion in a beam sea will be significant, it reaches a situation of dynamic stability. The corners of barges are subject to heavy impacts during operations; thus they must be heavily reinforced. Fenders should be provided on the corners to minimize impact damage to other craft and structures. Fenders should be provided along the sides to minimize damage to the barge itself from other boats and barges as they are docked. These may be a combination of integral fender strakes plus renewable fenders. Bitts are provided at the corners and at intervals along the sides to enable the securing of the barge and any other craft which come alongside. Towing bitts are provided on both bow and stern.

Consideration must be given to the need to temporarily weld padeyes to the deck in order to secure cargo for sea. These padeyes must distribute their load into the hull; they cannot develop proper strength by just welding to the deck plate. They will be subjected to impact loads in both tension and shear. In modern offshore barge design, special doubler plates are often affixed over the internal bulkheads so that padeyes may be attached along them. Low-hydrogen electrodes should be used. Alternatively, posts may be installed, running through the deck to be welded in shear to the internal bulkheads.

The deck is often protected by timbers to absorb the local impact and abrasion of the load. This is especially needed for barges which will carry rock which will be removed by clamshell or dragline bucket, or upon which a tracked crane or loader will operate. Manholes are provided in the deck for access to the inner compartments. These must be watertight. There should be a heavy coaming to protect the dogs or bolts which secure the manhole. Once again the warning must be made about entering inner compartments which have been closed for a long period. They are probably devoid of oxygen and must be thoroughly aerated before entry.

Marine barges are often intentionally grounded (beached) in order to load or unload cargo. The beaching areas must be well levelled and all boulders and even large cobbles removed in order to avoid holing or severely denting the barge. Once beached, the barge should be ballasted down so that it will not be subjected to repeated raising and banging down under wave action at high tide.

When heavy loads are skidded on or off a barge, they punish the deck edge and side because of the concentrated loading. Skid beams are often arranged to partially distribute the load to interior bulkheads. A timber "softener" may be temporarily bolted to the deck edge. The barge must be analyzed structurally for each stage of loading to ensure that a side or bulkhead will not buckle under the temporary overload.

Cargo must be secured against movement under the action of the sea. Thus sea fastenings are designed to resist the static and dynamic forces developed under any combination of the six fundamental barge motions: roll, pitch, heave, yaw, sway, or surge. The dynamic component is due to the inertial forces which develop due to acceleration as the direction of motion changes. Roll accelerations are directly proportional to the transverse stiffness of the barge, which is measured by its metacentric height. Since barges typically have large metacentric heights, accelerations are severe. Conversely, if due to high cargo, the metacentric height is low, the period and amplitude of roll and the quasi-static force imparted by the load are greater, but the dynamic component may be less.

These loads are cyclic. Sea fastenings tend to work loose. Wire rope stretches; wedges and blocking fall out. Under repeated loads, fatigue may occur, especially at welds. Welds made at sea may be especially vulnerable because the surfaces may be wet or cold. Low-hydrogen electrodes will help. Chains are a preferred method for securing cargo for sea, since chain does not stretch. If structural posts are used, they should be run through the deck to be welded in shear to the internal bulkheads. The slot through the deck should then be seal-welded to prevent water in-leakage.

The effect of the accelerations is to increase the lateral loading exerted by the cargo due to the inclination of the barge by a factor of two or more. Flexing of the barge can also have a significant effect on support forces and the sea fastenings. Thus deeper and hence stiffer barges will experience a smaller range of loads than shallow, less stiff barges.

With important and valuable loads such as modules or jackets, sufficient freeboard should be provided to ensure stability even if one side compartment or end compartment of the barge has been flooded.

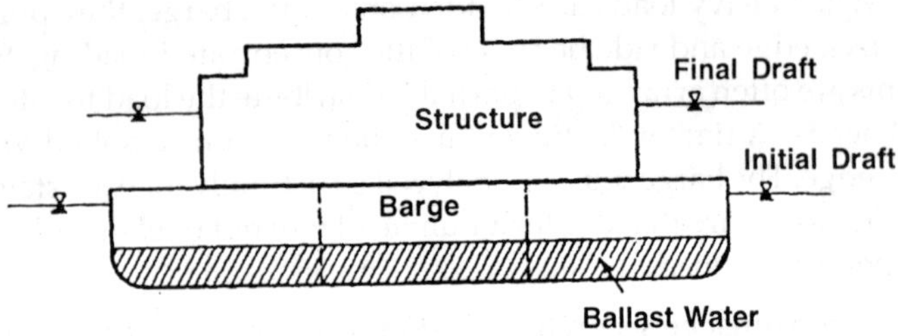


**Figure 2.1: Launching a Buoyant Structure from a Construction Barge**

Barges are normally designed to the standard loading criteria of the classification society. These criteria are usually based on submergence of the hull to the deck line, plus an arbitrary load of 3 m of water on deck. The typical barge is not designed for complete submergence.

Proposals are often made to build a structure on a barge, then submerge the barge by ballasting, and float the new structure off. Actually this has been successfully carried out in a number of cases: the construction of the pontoons of a drydock in northern Spain, the manufacture of several hundred shallow draft concrete hulls and posted barges near New Orleans and the construction of Arctic offshore caissons in Japan.

If a barge is to be seated, fully submerged, on an underwater embankment or on the seafloor in relatively shallow water, then it can be tipped down, one end first. Thus the beam of the barge and the inclined water plane provide stability at this stage. Then the barge end touches bottom. Now the barge may be fully submerged, gaining its stability from the end of the barge reacting against the bottom. This practice is normally limited to a water depth about one third that of the barge length.

Note that as the barge is tipped down, the transverse water plane area and moment of inertia is reduced to about one half of normal. Therefore, transverse instability can develop before the end touches the bottom, and the barge can roll. This acts to limit the depth of water suitable for such an operation. To recover the barge from the seafloor, the reverse procedure is followed, raising one end first.

A barge seated on a mud or clay seafloor develops a suction effect, consisting both of adhesion and a true suction due to differential water pressures. To break the barge loose requires that full hydrostatic water pressure be introduced under the bottom and that the adhesion of the clay to the barge be broken.

Extensive experiments by the Naval Civil Engineering Laboratory at Port Hueneme, California, and confirmed by practical experience in the Gulf of Mexico show that breaking loose can best be accomplished by sustained water flooding at a low pressure, less than the shear strength of the clay. Higher pressures will just create a piping through to the sea and prevent development of any pressure. Periods of several hours may be required to develop a fully equalized head of water under the structure.

As with ship salvage, a fully submerged barge must be given only limited positive buoyancy; otherwise it may break loose suddenly. If compressed air has been used to displace water in open compartments, the vessel will achieve additional buoyancy with every meter of rise due to the expansion of the air inside and may become uncontrollable. For all the above reasons, submergence of standard barges must be considered only to shallow depths.

## The Shear-legs Crane Barge

The term *crane barge* is used to denote an offshore barge equipped with a sheer-legs crane, hammerhead crane, or fully rotating crane. A shear-legs crane can pick loads and luff but not swing. The shear-legs consist of an A-frame made up of two heavy tubulars or trussed columns held back by heavy stays to the bow.

The shear-legs barge is maneuvered by deck engines, tugs, or mounted outboard engine propellers. The crane moves in to the side of the cargo barge, picks the load, then moves as necessary to set the load in exact position. Modern torque-converter deck engines and propellers with variable pitch allow a high degree of accuracy in positioning to be obtained, for example, of the order of 50 mm. One of the advantages of a shear-legs crane barge over a fully revolving derrick barge is that the load is always picked over the stern end, hence preventing list from the swing of the crane.

The sheer-legs crane is also much less costly than a full-revolving crane, both in first cost and in maintenance. Because of the need to move the entire barge to proper position to set the load, its operations are slower than those of a derrick barge. Further, it cannot choose its heading to minimize motion response to the sea. A sheer-legs crane barge is normally capable of ballasting down by the bow, to offset the trim induced by picking of the load. The barge must, of course, be designed to resist the hogging moment which then occurs when the load is picked.

The ability of a sheer-legs barge to lift a module or other large spatial load to a height (for example, in order to set it on a platform deck) is limited by the necessary length of slings and by the interference between load and the shear-legs themselves. The load cannot be allowed to swing into the shear-legs or it may buckle them. Swinging of the load due to pitch will, of course, increase this danger. To prevent such fore and aft swing, tag lines should be used to suck the load slightly in toward the stern; gravity will then prevent it from swinging in this direction. Swinging transversely can be snubbed by the use of tag lines as well.

Typical tag lines for offshore crane barges are 1/2 to 5/8 in., 6 × 37 wire lines to ensure flexibility, and are controlled by air or hydraulic hoists. Care must be taken to prevent their chafing as the load is moved to new positions in three-dimensional space. Softeners should be provided as necessary.

To pick loads from a barge at sea and then set them on a platform, the shear-legs are usually fixed at the appropriate orientation to serve both. Luffing of the shear-legs, that is, raising the shear-legs them selves, is awkward and slow and should normally be avoided. The load should be hoisted from the barge at the top of the heave (of the barge) so that 6 s later, on the next cycle of heave, the load will be clear of the barge. The operator (and foreman) will watch and try to catch a relatively higher wave on which to start the pick. Hoisting speeds depend on the number of parts of line in the blocks and, of course, on the rated speed of the engine and the amount of wire on the drum.

When it comes to setting the load, the problem is reversed. The load will tend to first make contact while the barge is near the bottom of the heave cycle; 3 s later, before the hoist engine can overhaul to slack the lines, the crane barge may lift the load up again. Under any significant sea state and pitch response, the load becomes a battering ram. Therefore, the crane barge should be fitted with a free overhaul capability to allow the load to remain seated once it has landed. In

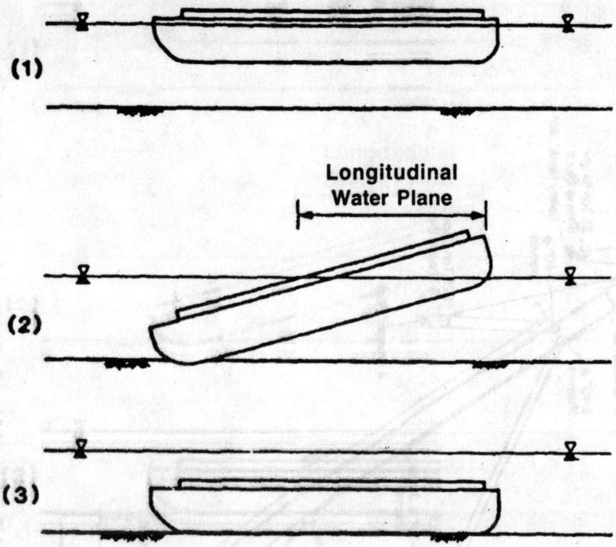


**Figure 2.2: Tipping a Barge to the Seafloor in Shallow Water**

any event, the skillful operator will try and set the load during a period of minimum motion and as close to the top of the crane barge's heave cycle as practicable, to give time for overhaul.

The slings used to lift typical modules and other heavy loads are very heavy and awkward. A whip line, single-part, is run over a sheave at the boom head to help lift the eye of each leg of the slings over the hook.

The deck engines of a sheer-legs crane barge must be adequate to control the barge's motion in yaw, sway, and surge to a very close tolerance despite the state of the sea. This requires an excess of power as well as torque-converter controls or equivalent. Fairleads must be carefully laid out to ensure a proper fleet angle from the winch and to ensure that they will properly follow the changing position of the barge.

Shear-legs crane barges were used to set the 200-ton precast concrete dome and breakwater segments of the Ninian Central platform. Shear-legs crane barges have been used to set the 3000 ton

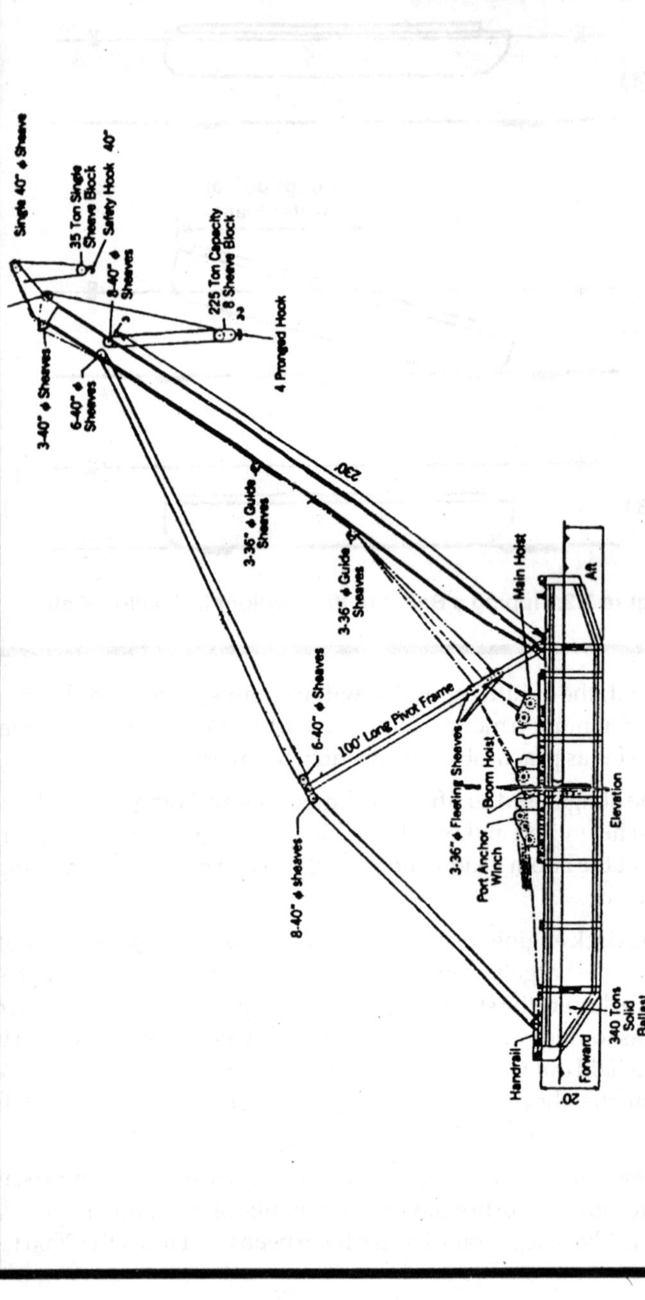


**Figure 2.3: Crane Barge Used to Construct Hay Point Terminal No. 2, Queensland, Australia; 225-ton Capacity at 20 m Over Stern**

approach caissons of the Great Belt Eastern Bridge and those of the Øresund bridge. Three crane barges, rigidly positioned relative to each other by multiple lines, were used to lift the 1200-ton quarters modules onto the Statfjord A platform. This is an inherently hazardous operation, but was carried out successfully because of the provision of oversize lines, adequate deck engines interconnected to a single control station on the central barge, and rigidly tying the three barges together to prevent relative movement.

A crane barge was successfully used on the Port Latta, Tasmania offshore terminal to set jackets into pre-installed frames with a tolerance of only 50 mm. A job-built shear-legs crane barge was used to set the superstructure of both this terminal and a similar iron-ore terminal in Queensland. Thus the capability of a crane barge for extensive use offshore should not be eclipsed by the currently more popular but much more expensive fully revolving offshore derrick barges.

Hammerhead crane barges have fully fixed hammerhead cranes. They operate in the same way as shear-legs crane barges, but cannot luff. The Svanen has a capacity of 8000 tons. It was used to get the piers, shafts and girders on the Great Belt Western Bridge, the Prince Edward Island Bridge, and the Øresund Bridge.

## The Damage for Marine Construction Vessels

Marine construction vessels are subjected to collision from barges and boats to a far greater degree than normal vessels engaged in transport. The latter avoids close proximity to other vessels, whereas an offshore construction vessel must work with these other craft alongside. The marine construction vessel is frequently picking and setting anchors; it is not unusual for the fluke to rip into the side. Finally, this rig must also work adjacent to platforms and other structures. Every precaution is taken to avoid collision with the structure because of the danger to the equipment, facilities, and wells; for an operating platform or terminal there is also the danger of fire from hydrocarbon release.

Damage control considerations require that vulnerable areas be subdivided into smaller compartments that all manholes and most doors be equipped with watertight gaskets and dogs, so that they may be kept closed except when actually in use, and that areas where anchors will rub or boats lie alongside be armored or fendered as appropriate.

During tow or when moored in heavy weather, green water will come over the decks of most barge-type vessels. An inadequately closed manhole will let in a large amount of water within a short time.

Temporary attachments and supports are frequently welded to the deck. If welded only to the deck plating, they may pull free; the welds are in tension normal to the deck plate, and the deck plate may be unsupported below that point. Therefore, holes are frequently cut in the deck plating so that attachments may be welded in shear to the bulkheads below. These must subsequently be seal-welded to the deck plate to prevent water entry. These temporary attachments, especially padeyes, winch foundations, and mooring attachments, are often subject to extreme lateral impact loading. The connection should be detailed so that failure occurs in the connection, not in the vessel's structure.

Construction vessels often take a sudden list due to a shift in the load, for example, as a crane swings or a heavy deck module is moved to one side or lifted off. These sudden lists may coincide with a roll and temporarily submerge an above-deck door, which has been left open, or a vent, or other opening below the oncoming wave. Other flooding accidents have occurred due to broken portholes (or ports left open).

Workboats and small barges are often pulling heavy mooring lines, whose weight may cause a temporary trim down by the stern or bow, resulting in wave overtopping. Boats pulled or running astern are especially subject to taking water over the stern sheets. Flooding into the stern well may have several severe consequences. It may enter the engine room or control room spaces and short out the power. Even a small amount of water in a compartment gives a free surface which reduces the metacentric height; that is, the righting moment available at the waterline of the vessel is reduced by the free-surface effect of the partially flooded compartments.

Watertight closures designated to be closed during operations must be kept closed and dogged. Two serious accidents occurred to workboats during the 1970s when, because of the calm seas and warm weather, engine room doors were left open for ventilation. An operational event caused water to be shipped over the deck; this in turn flooded the engine room spaces. Both boats sank rapidly, in each case with serious loss of life.

In a somewhat similar case, in dead calm weather, a semisubmersible derrick barge was making a heavy lift. The load swung to one side, the vessel heeled, and the swing engines and brakes were unable to hold the crane, which rotated to the beam, causing a very extreme heel, so that the upper deck was awash. Doors on the upper deck, supposed to be closed at all times, were wide open. In this case the alert crane operator prevented a catastrophe by lowering the load into the sea.

Construction vessels are usually equipped with ballast tanks to enable the list (heel) or trim to be controlled. These ballast tanks typically have vents which extend in a gooseneck above deck and are equipped with a flame-arreshing screen and a flap value. One purpose of these vents is to prevent accidential overpressurization of the tank, which might rupture a bulkhead and flood an adjacent space. However, these vents become plugged or may even be intentionally blocked off, for example, in order to store cargo on the deck space in question. The result may be overpressurization and a rupture of an internal bulkhead.

The more sophisticated offshore equipment of today–derrick barges, pipe-laying barges, launch barges, and semisubmersibles–have complex ballast systems to enable their list and trim to be rapidly controlled, even as operations are being carried out. Accidents, even capsizing, have occurred when the controls were short-circuited. Therefore, emergency manual controls are also provided. The crew must be trained regarding procedures after a malfunction.

Valve stems sometimes break loose, so that they appear "closed" when actually the valve gate itself is still partially open. Critical valves should be equipped with remote indicators. Critical valves have been opened for testing and then inadvertently failed to be closed afterward. These should be equipped with locks or tags as appropriate.

Steel working under cyclic or impact loads can be subject to fracture, especially at low temperatures when it drops below the transition value. Usually these cracks start and propagate with repeated cycles. Careful inspection of critical areas can locate these incipient cracks before they have propagated to a dangerous degree. They can then be repaired or a crack-arresting hole drilled or strap installed. Decks of barges are especially vulnerable since they are exposed to low temperatures and high stresses.

Closed compartments inside steel vessels can be very dangerous to human life. The steel corrodes slowly but continuously, using up the oxygen in the compartment. In other cases, heavier-than-air gases such as carbon monoxide may accumulate in the lower areas and bilges. Therefore, all compartments which have been closed and all tanks should be thoroughly ventilated before entering.

Fire aboard a vessel at sea is one of the traditional worries. For many fires, the most effective way of fighting it is to close off all air supply and cool down the adjoining bulkheads and decks by water spray. Fires can jump across steel bulkheads by igniting the paint on the other side. Electrical fires and hydrocarbon fires should not be fought with water. Acetylene tanks must be chained or strapped tight to prevent "falling, fracture, and ignition."

Materials, such as casing and all separate units stored on the deck of barges, must be secured against displacement in the event of a sudden list. Casing and pipeline pipe are especially dangerous because of their ability to roll and the large tonnages involved. Shifting loads can cause the vessel to capsize. Steel sheet piles on the deck of an inland marine barge have shifted as the crane barge listed while picking a load.

Often a large module is transferred from a cargo barge or a shore base to sit on the deck of a derrick barge. Even though the duration of exposure is relatively short, the module should be promptly secured with chains or wire lines so that it cannot shift even if the derrick barge rolls.

All lifesaving equipment should be maintained in full operating condition at all times. When a capsule or raft or firefighting gear must be removed in order to carry on operations, it must be relocated or reinstalled immediately afterward. Emergencies can occur at any time and, according to "Murphy's second law," will occur at the worst possible time.

## Different Types of Barges

### Fully Revolving Derrick Barges

Fully revolving derrick barges are the workhorse of offshore construction. As with the shear-legs crane barges, they are fitted with deck engines and full mooring capability, only here the emphasis is on stabilizing the barge's position rather than close control in positioning, since the derrick barge normally remains stationary during any particular operation.

The typical inland marine derrick barge has a capacity of 50 to 300 tons, whereas an offshore derrick barge has a crane capacity of 500 to 1500 tons. To handle ever larger modules and deck sections, capacities have been rapidly increased in recent years, with the latest offshore derrick barges having two cranes, each rated at 6500 metric tons each, or a total of 13,000 T.

The derrick barge represents a compromise (or optimization) of opposing demands. Structural and naval architectural considerations require it to be located forward of the stern a distance 20 to 25 per cent of the length, that is, at the one quarter or one fifth point. The barge should be wide enough to minimize list as the crane swings and to provide adequate distribution of the structural load.

On the other hand, the effective reach of the crane and its load capacity is diminished by the distance from the boom seat to the stern or side of the barge. One way to meet these two contrary demands is by the use of a large swing circle which moves the boom seat closer to the barge end while maintaining the center of rotation and support well back.

A major consideration is the list of the barge under fully loaded or no-load conditions. The counterweight is usually designed to limit the list under half load, hence under no-load the barge may list opposite to the boom. This list can be reduced during operations by booming down while swinging under no-load. The swinging is carried out by swing engines driving the bull wheel. Due to list, the crane is often forced to swing "uphill" under load. Offshore cranes are therefore provided with two and sometimes three swing engines. The list also places heavy structural loads on the crane tub, which forms the structural connection to the barge. Hence, its design must provide proper structural reinforcement for bending and to prevent buckling under inclined compression loads. Land cranes mounted on barges often fail as a result of collapse of the tub or center pin.

The advantages in operations of a fully revolving derrick are many: the ability to pick off a barge or boat alongside or even from the deck of the derrick barge itself, the close control of positioning to be able quickly to reach any point in three-dimensional space with one set of controls, the ability to follow the surge motions of a boat or barge alongside in order to pick a load from off it, and the ability to orient the derrick barge in the most favorable direction to minimize boom tip displacements and accelerations.

When setting large and heavy loads, it is the boom tip motions that control. These are affected by motions of the barge in each of the six degrees of freedom. When working far out over the stern, pitch amplitudes will be amplified. When working over the side, it is roll which causes the most difficulty. Computer programs have been developed to assist in selecting the proper heading, which treat the barge and load as a coupled system. A skillful barge superintendent and crane operator will take advantage of the "groupiness" of waves to perform a critical pick or setting operation during a succession of low waves.

As with a crane barge, tag lines must be used to control the swing of the load. As contrasted with sheer-legs crane barge, the position of the load relative to the barge is constantly changing; hence the tag line engines are fitted to the crane body and revolve with it.

A load suspended from a boom tip is a pendulum. While the load line length is usually too long for direct resonance, the load may tend to get dynamic amplification from lower-frequency energy. The practical solution is to raise or lower the load quickly through those positions which develop amplified response.

Marine cranes are usually designed to work under their rated loads up to a 3° list. The load capacity ratings for marine cranes are based on 2° roll at a period of 10 to 12 s, which equates to an acceleration of 0.07 *g*. The swinging of the load develops lateral forces on the boom. Hence, offshore crane booms are designed with a wide spread at the heel (usually 1/15 of boom length or more). This in turn means that the boom lacing (bracing) members will be subject to buckling; they must be properly designed to prevent this mode of failure. Booms today are made of high-strength steel, usually round or square tubulars. This makes them lighter and hence increases the effective load capacity of the crane and reduces the inertia in swing. However, it means that welds are more critical and that buckling becomes a common mode of failure. Good design and fabrication will take care of these. It also means that the boom is much more sensitive to lateral impact from the load itself or to failure under an accidental lateral loading. It means that attachments such as padeyes for snatch blocks and so forth must be affixed to the boom only after careful engineering and with fully controlled welding procedures suitable to the grades of steel involved.

One of the potential hazards with offshore derrick barge operation is that, although the lifts have been carefully engineered for load and reach, in the actual situation the derrick barge surges farther away from the platform and moves laterally. The operator, intent on the load and the landing site, booms out and swings beyond the crane's capacity. This may result in a direct failure of the boom or may result in a loss of swing control which accelerates as the barge lists. Offshore derrick barge cranes are fitted with automatic warnings to alert the operator when allowable load-radius combinations are being exceeded, but swing control is normally a matter of judgment.

To snatch a light load out of a supply boat, a single line, the whip, is preferred. It can raise the load fast enough to prevent an impact on the subsequent heave cycle. Raising a heavy load from a barge is more difficult since there may be 24 or more parts in the line and the barge will rise as the load is lifted, increasing the risk of impact of load and barge deck.

A similar problem occurs when setting a heavy load. When setting on a platform, the deck will usually be above the sight lines of the crane operator, the operator is working blind, dependent on signals. Hence, one or more guiding devices are needed. Tag lines from the crane barge may bend over the edge of the platform deck; if they chafe, they may part at the worst possible time. Softeners should be provided. Structural guides may be preinstalled on the platform so that the load, once set within 0.5 m or so of position, automatically guides down to the correct location. These guides must have sufficient height so that the load does not ride up out of them on the next pitch-heave cycle. If that were to happen, they could puncture the load rather than guide it. Taut guide lines can be employed to help pull the load to the correct position. A system of guides that often works well is to use two columnar guide posts. Suspended loosely from the load are two pipe sleeves of larger diameter. These can be hand-fitted over the posts; when the load is lowered, the sleeves will guide the load into place. Alternatively, loosely hanging pins (smaller-diameter pipe) may be entered into the tubular posts. Tag lines and winches may be installed on the platform to assist in guiding the load into place. Another solution is to set the load only to an approximate location, landing it on softeners such as timber or rubber fender units or used earthmover tires. After it has been landed in approximate position, it can be skidded to final exact position by hydraulic jacking equipment of the type commonly employed on oil-drilling rigs. This is the procedure often adopted when setting trusses or other unwieldy superstructure elements.

There is an arbitrary functional division that exists between offshore construction crews and offshore drilling crews. Neither seems to fully appreciate the problems of the other. This has resulted in much needless work and several accidents. Close coordination and communication are essential.

The lower (traveling) block and hook of a large offshore derrick can weigh 20 to 30 tons or more. As it is brought up close to the boom housing, it may get into resonance with the roll of the barge. A special hook control tag line is required. The traveling block-hook combination should never be left hanging at short scope. A sea may come up that excites the hook and makes it impossible to secure. Thus, except when the crane is being used, the block should always be fully stowed and the boom lowered into the boom cradle and secured. This will also reduce fatigue wear on the swing gear.

When a derrick barge is working alongside a platform, the moorings are laid out in a pattern which allows the barge to reorient and relocate as necessary to reach as many parts of the platform as possible. Care must be taken that during a reorientation, the mooring lines are not allowed to cross one another. Although there are exceptions, as a general rule, mooring lines should never cross; it prevents retrieval of the underneath line, and it may lead to erratic reactions from the lines as the load in one changes its catenary and affects the other. Worst of all, one line may snag the anchor of the other line.

## Catamaran Barges

For heavy lifts in harbors and on rivers, especially for submerged prefabricated tunnel (tube) segments, catamaran heavy-lift barges are frequently employed. These consist of two long barges, spread apart, and joined over-the-top by gantries. Often, but not always, the gantry legs are pin-connected at the centerlines of the barges, so as not to impose any listing moment on the barges, thus necessitating additional ballast and consequent increased draft. This also allows the barges to undergo a small degree of roll independently without affecting the gantry trusses.

Gantries are usually equipped with twin gantry trusses, one at each end of the barges, to enable lifting and/or lowering of long prefabricated segments such as tubes. Two lifting devices are arranged on each truss, making a total of four lift points for the system. To resist fore-and-aft differential movement of the two barge hulls, horizontal connecting trusses are installed at one or both ends. Catamaran barges are also used for screeding of underwater foundations. To maintain exact grade and to be unaffected by waves and swells, they may use the semisubmersible concept in which the hulls are ballasted down below water, with columns or shafts extending up to support the superstructure. To offset the effect of tidal changes in water elevation, precast concrete clump weights may be lowered to the bottom to maintain a constant elevation, offsetting the increased buoyancy of the rising water on the columns.

## The Jack-up Barge

The jack-up barge has proved to be a very useful construction "tool," especially when working in turbulent sea areas, or breaking waves such as shoal or coastal waters, and in swift currents. Where a great many operations must be carried out at one location–for example, at an offshore terminal or bridge pier–the jack-up construction barge is especially valuable. The barge is outfitted with four to eight large jacks and legs, built either of tubulars or fabricated steel. The barge is towed to its work position and jacked up free of the waves to perform its work.

The typical sequence starts with the barge moving to the site with its legs raised. Upon arrival at the site, it is moored with a spread mooring. Construction jack-ups can operate only in relatively shallow water, 30 to 60 m, with 100 m as an extreme, so the use of a taut mooring is practicable.

With the sea state being calm (waves and swells must usually be less than 1 m), the legs are lowered to the seafloor and allowed to penetrate under their own weight. In some soils, penetration can be aided by jetting and vibration. Using the jacks on one leg at a time, the barge acting as the reaction, the legs are forced into the soil. With all legs well embedded, the barge is jacked up clear of the water. This is the most critical phase, since wave slap on the underside of the barge may cause impact loads on the jacks and may shift the barge laterally, bending the legs. To cushion the impact, special hydraulic cushioning may be connected to nitrogen-filled cylinders; alternatively, neoprene cushioning may be employed. Once well clear, the barge is raised up to its working height. Then the legs may be cut loose, one at a time, and a pile hammer used to gain even greater penetration. Since uneven settlements may take place as a result of time, operations, and wave energy input into the legs, the jacks have to be periodically reactivated to equalize the load at each. This is especially necessary during the first few days at a site.

To leave a site, the sea must again be calm, with waves and swells usually less than 1 m. The mooring lines are reattached, slack. The barge is then jacked down until it is afloat. Once again the critical period is when the waves are hitting the underside. The mooring lines are tightened. Then the legs are jacked free, one at a time. If legs do not pull out easily, several techniques can be applied. The fastest is jetting. In clays, a sustained load may eventually free the leg. Also in clays, water injection at low pressure to break the suction may be more useful than high-pressure jetting, which leads to the formation of escape channels. The same process may be used to free legs in sand, also at low sustained pressure. In no event should an attempt be made to free the legs by lateral working of the barge. A bent or jammed leg may result, with very serious consequences. In one case in Cook Inlet, Alaska, the leg was jammed by the high current working on the barge side during the refloating operation. Then the tide rose some 6 m, flooding out the jack-up barge itself.

High currents may create local eddies around the legs, leading to scour and loss of lateral capacity. Mats can be built onto the bottom of the legs, so that when the legs are jacked down they take their support from the seafloor. A short stub leg may penetrate below the mat to provide shear resistance against sliding. Since jack-up performance is so highly dependent on the seafloor soils, it is essential that a thorough geotechnical evaluation, including at least one boring, be made at each site. Of particular concern are layered soils, in which a leg may gain temporary support but then suddenly break through.

In clay soils, where jack-ups have previously worked around the site, holes will have been left which now may be partially empty or filled with loose sediments. If a leg is now seated adjacent to such a hole, it may kick over into it, losing both vertical and lateral support and bending the leg. A general rule of thumb is to plot the previous leg positions (if known) and to space the new leg locations 4 to 5 diameters away. This, of course, is another advantage of the mat-supported jack-up legs: the mats can span local anomalies.

Walking jack-ups have been built, varying in size from a small test-boring rig capable of walking through the surf to a monstrous dredge hull on jack-up legs. These rigs are equipped with two sets of legs (six or eight in all) supporting a double framed hull (or segments) so that it can successively launch forward, lower the legs, take its full support forward, pick up the legs behind, and retract the rear into the forward section. The rear set of legs is now lowered to give added support during operations and then to enable the forward set to be picked up once again. Such walking jack-ups eliminate the need for the hull to be lowered into the sea in order to move. The smaller walking jack-up rigs are especially useful for taking borings in the surf zone. Unfortunately, the large walking jack-up dredge proved too costly and slow, hence has been taken out of service. The smaller walking jack-up has proved quite successful.

Large jack-up construction rigs are most applicable where the sea conditions are highly variable, with frequent periods of calm, so that the rig may find convenient times to move. On the other hand, if numerous moves are required–as, for example, in laying outfall sewer pipes–then persistently rough seas may delay moves so long as to render the jack-up uneconomical. One disadvantage of the jack-up occurs during the transfer of loads from barges or supply boats. Here the jack-up concept again becomes weather sensitive, for the barges must not be allowed to contact the legs or they may damage them.

Jack-ups provide a fixed platform, free from motion response to the seas. Hence, they are ideal for carrying out operations such as grinding a rock foundation in order to seat a caisson, as was done on the Honshu-Shikoku Bridge (Koyama-Sakaide Route, Pier 7A). They are also ideal for screeding the foundation site.

Statistical studies covering both jack-up drilling rigs and jack-up construction rigs show that they have been six times more likely

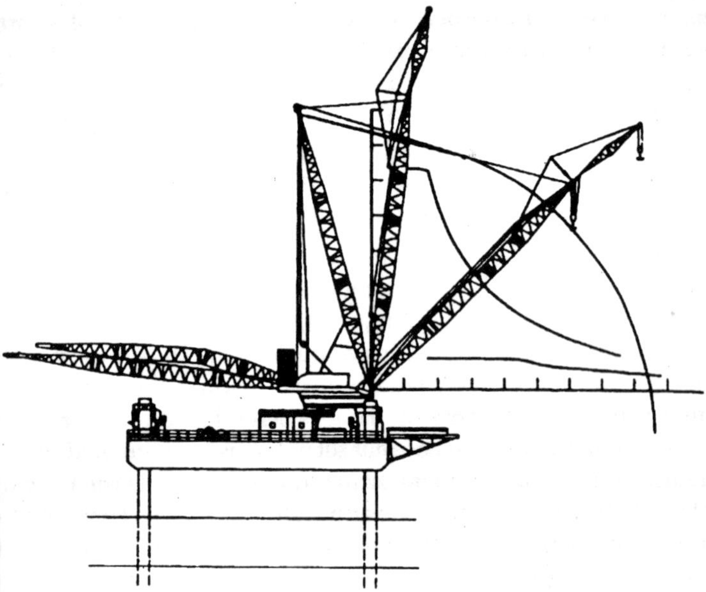

Figure 2.4: Jack-up Construction Derrick

to suffer serious damage or loss during relocation and transit than they are when on location. This is primarily due to the barge having its legs fully raised, thus creating a very high center of gravity. Some jack-ups therefore have telescoping legs.

As with the semisubmersible concept, the jack-up principle has been applied to smaller, special-purpose construction rigs tended by an offshore derrick. The derrick barge with its large mooring system can be used to position the jack-up and if necessary to help in penetrating its legs and later to help in retracting them. Meanwhile the jack-up rig forms a vertically stable work platform for such sensitive operations as coring and sampling operations or for submarine pipeline repairs.

The jack-up has been used to set heavy loads. In this case, a barge carrying the load is floated between the legs. The load is then lifted by direct hoisting from the jack-up deck above, the barge removed, and the load lowered to the seafloor. This operation, with a barge between the legs, is obviously highly hazardous and should only be attempted under ideal sea conditions, with adequate lateral controls to ensure that the barge cannot hit the legs. This concept was used to set the 600-ton precast concrete caissons for the Columbia River (Oregon) bridge at Astoria.

## Pipe-laying Barges

The pipe-laying barge is a highly sophisticated vessel which constitutes the key element in an offshore submarine pipeline installation system.

The functions of the barge are to receive and store pipe lengths, assemble and weld them into a single length, coat the joints, and lay the pipeline over the stern to the seafloor. Operations involved in accomplishing the above include:

1. Positioning the barge
2. Handling pipe lengths from a barge or supply boat to the barge deck
3. Double-ending (optional)
4. Lining up and completing the initial hot pass weld
5. Completing the welds
6. X-ray
7. Applying tension to the pipeline
8. Coating the joints
9. Laying the line out over the stern, usually by means of a stinger
10. Moving the barge ahead on its anchors
11. Shifting anchors continuously ahead
12. Recording positions of laid pipe accurately
13. Radio communications to boats, shore, and aircraft
14. Helicopter and crew boat personnel transfer
15. When weather conditions dictate, "abandoning" pipeline onto the seafloor in an undamaged, unflooded condition
16. "Recovering" an "abandoned" line and recommencing pipe-laying operations
17. Davits to permit supporting a section of the line uniformly for riser tie-in or repair
18. Diving support for inspection
19. Housing and feeding of up to 300 people

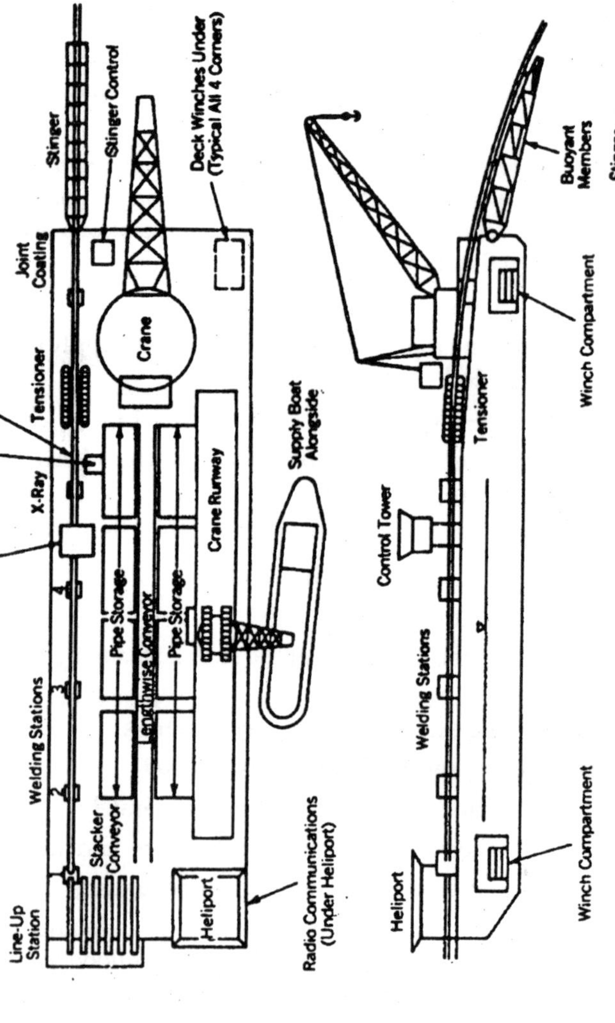


**Figure 2.5: Second-generation Pipeline Lay Barge**

In the above listing, the word *abandonment* is to be construed as a temporary cessation of work and laydown due to the real or threatened onset of a storm.

Such a long list of requirements inevitably requires a large offshore barge. Both heavy-duty standard offshore barges and semisubmersible hulls have been used. The length of the barge is further dictated by the number of welding stations required in order to maintain the desired rate of progress. Since deep-water pipelines inevitably have thick walls, many passes are needed in order to complete full-penetration welds. The more stations there are, the less time needs to be spent at any one station, and hence the ability to increase the lay rate.

To move the barge ahead requires many mooring or anchor lines. Large two and three drum waterfall winches are mounted along the sides of the barge. The mooring lines lead from the winches over direction-changing sheaves to submerged fairleads and thence to the anchors.

To handle the pipe lengths onto the pipe-laying barge, a large crawler crane usually is used, one which can quickly snatch a 40-ft length from a tossing supply boat or barge at the top of the heave cycle.

A number of heave-compensating devices have been tried, with a signal line from the hook to the boat, for example; however, the snatch method still seems most effective. Once the pipe is stored on board, the next operation may be double-jointing. This usually does not speed up the overall pipe laying, but does reduce the number of specially skilled welders required.

The pipe length, single or double, is then conveyed end-0 and sideways to the lineup station. The pipe rolls onto the rack, which is hydraulically controlled to line up and position the pipe accurately. An internal lineup clamp is applied to join the new section to the previous one so that the first "hot pass" weld may be made. The joint then moves iteratively to the several welding stations, where the weld is chipped and cleaned and new metal deposited.

The weld, once completed, moves to the X-ray station, where pictures are taken, reviewed, and approved. In the case of a reject, a cutout must be made and the removed weld rewelded and reinspected. Aft of the X-ray station, the tensioner is installed.

Tensioners are usually of the caterpillar track type, using polyurethane tracks pushed tight against the rough coating by multiple hydraulic jacks. The tensioning force is thus applied to the pipeline by friction. At the next station, the joint is coated with bitumastic.

The pipeline is now ready to move down an inclined ramp and out over the stinger.

Early (first-generation) stingers were long, hinged ladders, partially buoyant, not unlike a dredge ladder in concept. They in effect formed a ramp down which the pipeline ran to the seafloor, with minimum bending stress. Wheels or rollers were provided to reduce friction and to prevent abrasion of the coating.

Second-generation stingers were articulated to accommodate the higher-frequency wave motions to reduce the stress in the pipe. These stingers were also buoyant, some even employing the semisubmersible or spar principle to minimize heave response to the waves.

With the development of improved tensioners has come the third-generation stinger, a curved cantilevered ramp, supported on the barge. It guides the overbend of the pipeline down to its point of departure. The fourth-generation pipe-lay barge uses a high angle for assembly and the J-lay procedure.

Early welding lines, ramps, and stingers were also put on one side of the barge, usually the starboard side, originally as an appendage to an offshore derrick barge. With higher tensioner forces, the tensions in the anchor lines leading forward became critical. The most recent pipe-laying barges therefore have the welding line and stinger on the centerline of the vessel.

Control of the pipe on the stinger and consequent control of the tensioner force require the use of load cells or similar devices on the stinger so that the pipe reactions and point of departure may be read out in the control room.

For abandonment and subsequent recovery, a large, constant-tension winch is required, positioned so that it can lead its line down the pipe-laying alignment.

Finally, there must be provided all the housing, feeding, and support functions: cabins, mess room, recreation hall, machine shop, power generation, pumps, and winches.

A large crane is on the stern. The original cranes were there to enable the pipe-laying barge to also double as a derrick barge. However, a long-boom crane capacity is also needed for setting risers and for installing and removing the stinger.

## Towboats and Crew Boats

Towboats are of several basic types. The large ocean-going, long-distance towboat is capable of operating 20 to 30 days without refueling. It is designed to move to any part of the world to carry out a major towing job. Such vessels may be up to 80 m or more in length and carry a crew of 16 to 20. They can run light at speeds of 12 to 15 knots. Harbor and other inland towboats are smaller and more maneuverable.

Towboats are often described in terms of horsepower, but this can be misleading. Indicated horsepower (IHP) measures the work done at the cylinders of the engine. Shaft horsepower (SHP) is the work actually delivered to the propeller shaft and may be 15 to 20 per cent less than IHP. Long-distance towboats typically have IHP ratings of 4000 to 22,000 HP.

Bollard pull, a much more meaningful measure, is the force exerted by the boat running full ahead while secured by a long line to a stationary bollard; that is, the boat is making no headway through the water. A rough relationship exists between IHP and bollard pull: a 10,000 IHP boat can exert 100 to 140 tons of static bollard pull. However, the relationship varies with the size of the propeller(s), whether single- or double-screw, and the draft of the towboat. The effective bollard pull falls off as the speed through the water increases. The largest tugs have bollard pulls of over 300 tons.

Large ocean-going towboats are fitted with the latest in navigational equipment: GPS, Loran C, radar, electronic positioning, and sonar. They can communicate by voice radio anywhere in the world. These boats may be fitted with a towing engine to enable them to maintain a constant tension on the towline, despite the varying response of the boat to the waves. Other operators prefer to rely on the use of a long catenary, adjusted during tow to span a full wavelength or more. Boat length should be 11 or more times the expected maximum $H_5$ for safe and efficient operation. In major storms, the boat may have to cut loose, in the hope that it will recover its tow after the storm has passed.

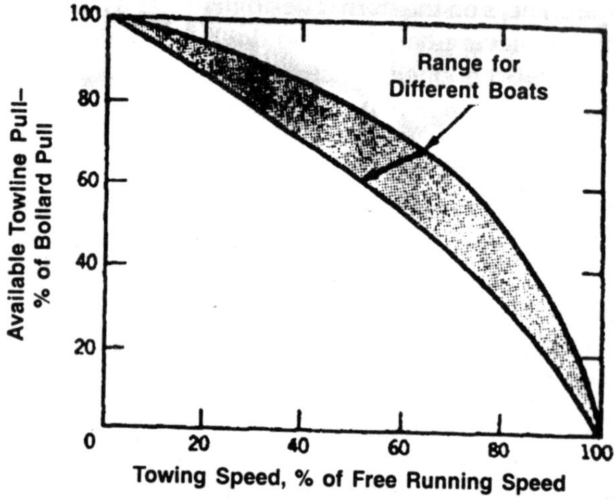


**Figure 2.6: Effect of Speed on Tug Towline Pull**

Shorter in length but still powerful are the boats designed for general operations in a specific theater of operations such as the North Sea. These boats are highly maneuverable, often fitted with a variable-pitch propeller that enables them to keep the engine running at full speed during critical positioning operations. They are usually equipped with bow thrusters to enable them to turn up into the wind while making no headway.

Ocean towboats range from 4000 HP for use in moderate seas to 11,000 HP for all weather ocean tows to 22,000 HP for towing the largest offshore platforms. Up to eight large boats have been used in tandem to tow a platform displacing 600,000 tons. Inland marine towboats range from 200 to 1000 IHP.

Now let us try to gather some details about crew boats.

Crew boats are used to transfer personnel from shore to offshore operations wherever sea conditions permit this to be carried out in a reasonable and practical manner. Crew boats are seldom used in the North Sea; distances are too great and weather conditions unpredictable. Helicopters are used instead. Crew boats are used in the Gulf of Mexico and offshore Southern California. Economics dictate that the boat should have as high speed as practicable. For non-planning boats, the required horsepower is proportional to the square of the velocity. Consideration has to be given to the boat's motions en route: one does not want the entire crew change to arrive seasick. Generally speaking, accelerations should be minimized by adopting a boat with as low a metacentric height as is consistent with safety. A high *GM* means a quick roll response and physical discomfort to the passengers. A boat may get into pitch reasonance with head-on or nearly head-on seas. This may be modified by changing the speed or heading or both. If the boat's length exceeds the wavelength, pitch response is reduced; however, this is usually only practicable in the Gulf of Mexico, not in the Pacific or North Atlantic, with their longer waves.

*Chapter 3*

# *The Offshore and Sea Environmental Problems*

## Introduction

The sea environmental problems can be studied as follows:

1. Those associated with sea and weather,
2. Those associated with oil and gas, and
3. Those due to mechanical shock and vibration and the structural limitations of the installation.

The salt-laden air and salt spray mists produced during bad weather have a searching corrosive effect on unprotected metal, as electrolytic cells are set up at these locations. Therefore any protective coatings used on metal equipment offshore should remain impermeable, and preferably contain a sacrificial metal such as zinc.

A variety of different enclosure and support structure materials have been tried offshore, with varying results, but experience suggests that the following materials are suitable:

### Stainless Steel

This tends to be an expensive solution, and needs to be carefully specified so as to avoid problems such as susceptibility to cracking. Stainless steel cable ladder and tray is more difficult to manufacture than that fabricated from the more common structural steels, and may be delivered with some deformity. This should be checked by normal quality control procedures, as even a small out-of-true error will make the installer's life very difficult. Stress corrosion in stainless steel takes place in the presence of chlorides, and related failures occur well below the normal tensile strength of the metal involved. The stresses during the deformation processes involved in the manufacture may remain locked up until 'season cracking' occurs in the presence of sodium chloride (salt spray). However, good quality low-temperature-annealed stainless steel products will provide long-term resistance to corrosion and are highly resistant to incidental damage.

## Grey Cast Iron

In grey cast iron, most of the carbon is present in the form of graphite flakes, which make the material softer, more machinable and less brittle than white cast iron. As the name suggests, cast iron is very fluid when molten and is therefore suitable for the manufacture of intricate castings. Its main use offshore is in the construction of flameproof enclosures. Its resistance to corrosion appears to be quite variable. Where it is exposed to salt spray, for example in flameproof control stations on lower-deck handrails, a galvanized finish is advisable. The variation in the effects of corrosion is probably related to the method of producing cast iron, which involves the remelting of pig iron in cupolas. The qualities of the cast iron produced will depend on the selection of the pig iron, on the melting conditions in the cupola, and on special alloying additions.

## Hot Dipped Galvanized Steel

This material is by far the most common for use in cable support systems offshore. The heavy duty grade should provide a service life in excess of 20 years, particularly if Corten A steel is used.

## Polycarbonate

This is a very tough plastic material used for junction box and similar electrical enclosures because of its corrosion-free property. It is particularly effective in areas close to the sea, where salt spray is common. It is resistant to mechanical damage and will deflect rather than break in most situations, but a heavy blow from a scaffolding pole, for example, is more likely to damage a polycarbonate enclosure than an equivalent steel one. Polycarbonate will tend to deform at elevated temperatures; it must therefore be shielded from the heat produced by flare stacks etc., and must not be used with equipment which has to operate during fire fighting operations.

### Manganese, Bronze and Gunmetal

These very heavy corrosion resistant metals are sometimes used with success for the casings of floodlights and similar exposed electrical equipment.

### Welded and Cast Structural Steel

Rotating machinery packages of all types are generally constructed of this material. It is not practicable to galvanize the whole package and therefore, as with the structural steel of the module or platform jacket, a suitable offshore paint system must be applied by the fabricator. The integrity of this paint system must be preserved during the equipment's transit, installation and commissioning, if the package is to be presented to the operator in good condition.

### Glass Fibre Reinforced Plastic

GRP has been used successfully offshore for a number of years. It may be used in electrical equipment in the form of small junction boxes and for cable ladder and tray. Although fire resistant, it will burn when subjected to a gas flame, but is more resistant to deformation, melting and fire than polycarbonate. Being strong and light and unaffected by sea water, it is useful particularly where installed in salt water spray conditions. However, as it will burn, long unbroken vertical runs should be avoided.

## The Equipment on an Offshore Installation

Much of the equipment on an offshore installation will be located in an area classified as potentially hazardous, because of the risk that flammable gases or vapours may be present in the atmosphere and could be ignited by equipment which creates electrical sparks or, during and perhaps for some time after operation, has an enclosure with a high enough surface temperature. This subject is covered exhaustively in a multitude of standards and codes of practice, many of which are listed in the Bibliography, and it is intended only to summarize the subject here. Those unfamiliar with the subject are recommended to study BS 5345 'Code of practice for the selection, installation and maintenance of electrical apparatus for use in potentially explosive atmospheres.'

An explosive atmosphere is one where a mixture of air and flammable substances in the form of gas, vapour or mist exists in such proportions that it can be exploded by excessive temperature, arcs or sparks.

The degree of danger varies with the probability of the presence of gas from location to location, and so hazardous areas are classified into three zones as follows:

**Zone 0 (more than 100 hours per year)**

In which the explosive gas mixture is continuously present or present for long periods. The conditions in such zones are usually regarded as too dangerous for any electrical equipment to be located in.

**Zone 1 (1 to 100 hours per year)**

In which an explosive gas/air mixture is likely to occur in normal operation.

**Zone 2 (less than 1 hour per year but more than 1 hour per 100 years)**

In which an explosive gas/air mixture is not likely to occur in normal operation and, if it occurs, will exist for only a short time.

By implication, an area that is not classified as zone 0, 1 or 2 is deemed to be a non-hazardous area.

The numerical values for exposure time shown above are for guidance only.

Now let us learn few more things about ignition temperature and flash point temperature.

The minimum temperature at which a gas, vapour or mist ignites spontaneously at atmospheric pressure is known as the ignition temperature. To avoid the risk of explosion, the surface temperature of the equipment must always remain below the ignition temperature of the explosive mixture.

Maximum permissible surface temperature are classified as in Table 3.1. Classification applies to the equipment, not the gas; the actual classification designation can be taken as the next below the ignition temperature of the gas.

**Table 3.1: Surface Temperature Classes**

| *Temperature Class* | *Maximum Surface Temperature (°C)* |
|---|---|
| T1 | 450 |
| T2 | 300 |
| T3 | 200 |
| T4 | 135 |
| T5 | 100 |
| T6 | 85 |

Motors to European standards conforming to EN 50 014 and 018 have temperatures 10°C lower than that shown in the table for classes T1 and T2, and 5°C lower for class T3 and below.

The flashpoint of a liquid or solid is the minimum temperature at which the vapour above the material just ignites by application of an external flame or spark, in standard test conditions.

The flashpoint gives a very useful indication as to how hazardous a material is, and is used when drawing up a schedule of hazardous sources for a particular installation. As discussed later in this chapter, equipment designed for use in hazardous areas should never allow sparks or flames to come in contact with the external environment, although in the case of Ex'd' equipment flames or sparks may occur within the enclosure. This is inherently the case with Ex'd' switchgear.

If the flashpoint of a substance is higher than 38°C, it is not normally regarded as a source of explosive hazard on North Sea installations. However, it should be noted that in hotter climates the storage area for the aviation kerosene (flashpoint 38°C) required for helicopters will have to be classified as hazardous, since there is a higher risk that ambient temperatures will exceed this temperature.

## The Degree of Ventilation

The degree of ventilation around the source of release will affect the classification of the hazard, unless one of the three following situations apply:

1. Open spaces with no structure or equipment restricting substantially free circulation of air, vertically and horizontally.
2. A module with a roof and not more than one side closed, free from obstruction to natural passage of air through it, vertically and horizontally.
3. Any enclosed or partly enclosed space provided with artificial ventilation to a degree equivalent to natural ventilation under low wind velocity conditions, and having adequate safeguards against failure of the ventilation equipment.

The number of air changes per hour required to meet the natural ventilation equivalent condition in the third case will vary to some extent, depending on the degree of hazard, but a figure of 40 changes per hour is quoted by underwriters. The positioning of fan inlets and exhaust outlets must be such that no unventilated pockets are left. The adequate safeguard is usually an alarm followed, after a suitable time delay, by the shutdown and venting to the flarestack of all process equipment within the module.

Process modules are usually ventilated such that the internal air pressure is slightly negative with respect to the outside, in order to retain any minor gas leaks. Safe area modules are kept at a slightly positive pressure in order to prevent the ingress of gas. The ventilation systems of different modules must be well segregated to prevent cross-contamination of leaks. Fire and explosion dampers are installed to automatically seal ventilation systems if fire or gas is detected. Ventilation intakes for both safe and hazardous areas should be located well within safe areas to minimize the risk of drawing gas into the ventilation system.

A common problem on older installations is the location of secondary pressurized control rooms in the middle of a hazardous module. A typical example of this is a gas compression module with its own control room. If the control room pressurization fails and remains failed for a timed period, the compression module should shut down and all non-certified electrical equipment in it should be isolated. Often, a loss of pressurization will only give an alarm; the decision to shut down the plant will be left with the operator, and prevention of gas ingress will depend on an airlock door system.

Apart from the serious ignition hazard that all the live non-certified equipment poses in this situation, the control operator's only means of escape is via the compression module where the gas leak has occurred. Owing to the amount of work involved in providing automatic isolation, with shutdown contactors for all the electrical supplies, and separating the instrument and control functions so that the main installation control room is unaffected, it could well be more beneficial to relocate the compressor control room to a safe area or possibly include its functions in an extended central installation control room. The last option would be more in line with the Cullen Report, as it would remove an extra manned control room. Both relocation options eliminate the need to provide an escape route through the compression module protected to some degree from fire and explosion.

A detailed discourse on the ventilation of offshore installations is beyond the scope of this book. However, the following is a list of basic criteria with which such ventilation systems should comply:

1. The positive pressurization of non-hazardous areas with respect to adjacent hazardous areas or the external atmosphere.
2. The containment, dilution and removal of potentially hazardous concentrations of explosive gaseous mixtures in hazardous modules, and the adequate segregation of hazardous area ventilation exhausts from ventilation and engine intakes.
3. The provision of comfortable environmental conditions in accommodation and normally manned non-hazardous areas, and acceptable working conditions in normally unmanned areas.
4. The provision of acceptable working conditions within hazardous modules.
5. The isolation of individual areas and control of ventilation in emergency conditions, in accordance with the shutdown logic of the installation's ESD, fire, gas and alarm systems.
6. The provision of combustion air for essential prime movers, ventilation air for escaping, fire fighting and rescue personnel, and purging air services required to operate effectively during an emergency.

To avoid depressurization and potential release of explosive mixtures, all exits to hazardous areas should be airlocked with self-closing doors. If single doors are used, a hazardous zone will extend outside the compartment.

Safe area control rooms may be located within hazardous modules provided they are kept pressurized with air obtained from a non-hazardous external area. Purging, alarm and timed shutdown facilities are required, similar to those for a pressurized motor. However, location of control rooms in safe areas is preferred for reasons given above.

## The Logic Flow for Selection of Hazardous Zones

It is common practice in offshore installations to require all motors for use in hazardous areas to be suitable for zone 1 areas, although in many cases the actual area classification might be zone 2. This allows areas to be reclassified from zone 2 to zone 1 without the need to replace motors; enables the safer transfer of motors from one hazardous area to another; and tends to rationalize spare parts holdings.

Where a motor is used as part of a variable speed drive system, the hazardous area certification will not be valid unless the motor has been tested for safe operation over the whole speed range through which it is required to operate. At low speeds the motor will have different heat dissipation characteristics and, if dependent for cooling on a rotor driven fan, will have a reduced cooling air flow rate. Therefore there is a risk that the enclosure temperature may exceed the required maximum surface temperature requirement.

It is now known that large motors whose casings are made up of bolted sections may suffer sparking across the section joints due to voltages produced by stray currents induced in the sections. To avoid risk of ignition, all such motors should now be fitted with copper braid bonding straps across each section joint in order to prevent any potential buildup across the joint.

In general, there are four methods of motor design to achieve suitability for use in hazardous areas, as follows.

The principle of protection, as with the metal gauze on the Davy lamp, is to control the flow of and to cool any burning gas escaping from the motor casing, so that any gas that has escaped is no longer hot enough to ignite external gas. The size of gap or flamepath

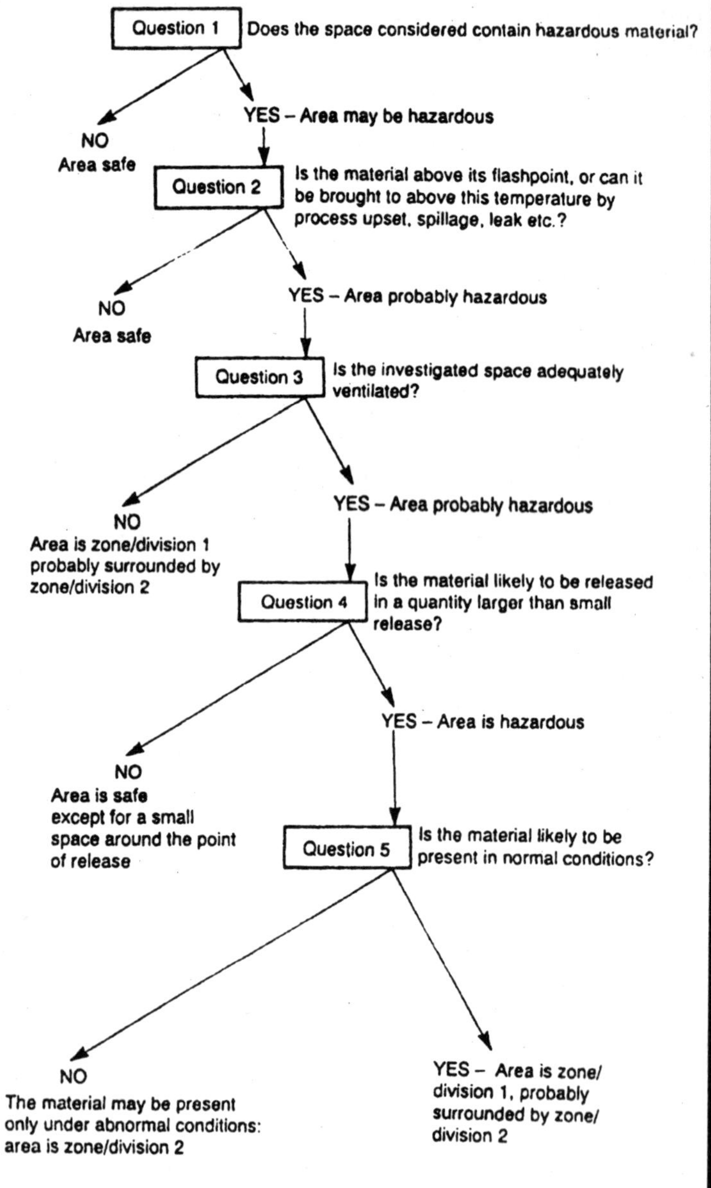


**Figure 3.1: Logic Flow of Selection of Hazardous Zones**

necessary to sufficiently cool the ignited gas on its way out of the enclosure varies according to the gas or vapour involved. Gases and vapours are subdivided according to experimental data which has been established to determine the maximum experimental safe gap (MESG). In the case of metal-to-metal joints in a flameproof motor, for example that of the frame to the end shield, these will consist of a long metal spigot fitting into a long recess which will normally be clamped tightly to fixing bolts. A flamepath will always exist between the shaft and the motor interior.

For safety, all the flamepaths or gaps in the motor enclosure must never exceed mandatory dimensions, and the casing of the motor must be strong enough to withstand an explosion caused by the ignition of the maximum free volume of air/gas mixture it can contain, with all flamepaths at minimum production values. It should be noted that this could be exceeded in very cold regions (polar latitudes) owing to the increased density of the gas, or at the elevated pressures used in diving. This type of protection is common on low-voltage motors up to a 315 frame size; on larger sizes, severe cost and weight penalties are likely owing to the cast method of construction.

## The Platform Structures

The platform structures installed in sea depths of 150 metres or more rotating machinery rated at 5MW or more may be operating near the top of a structure over 200 metres high.

The structure may resonate with vibrations produced by the machinery and, particularly if the machinery skid floor is cantilevered out beyond the main structure, a catastrophic failure may occur. To ensure that this is not the case, vibration analyses need to be carried out to establish structural resonance frequencies. If these are close to those of the machinery, modifications will need to be carried out in order to change the offending frequency to an acceptable value. At such heights, the swaying of the platform and the shock from drilling activities etc., may produce damaging torques on the bearings and shafts of large motors and generators if these are not catered for in the machine design.

An offshore production platform is a compact three-dimensional arrangement of modules. Some modules contain prime movers; others, such as separator and gas compression areas, require to exhaust ventilation air from hazardous areas; and all must take in fresh, uncontaminated air from the surrounding atmosphere. The prevention of cross-contamination between exhausts and intakes in all wind and weather conditions can be extremely difficult, and in some weather conditions it may be necessary to accept a small drop off in gas turbine performance due to the thermal contamination of other exhausts. Water curtains around turbine exhaust ducts are sometimes used to good effect in reducing contamination, by taking exhaust gas downwards, away from intakes. Prime movers for emergency equipment such as emergency generators and fire pump alternators are not so critical, because testing can be carried out in a favourable wind which blows the exhaust away from ventilation intakes. With emergency equipment, it is more important that aspects of emergency scenarios such as gas cloud boundaries and likely flame passages are given consideration.

On any platform, routine production maintenance and drilling operations in demand the movement of equipment, containers, scaffolding poles, drill-pipe etc. from the laydown area, at which it was received from the supply vessel, to its point of use, and then possibly the reverse journey back to the laydown areas prior to transfer back to the supply vessel. During these movements there is a risk that the item will be dropped on to, or swung into, some exposed piece of electrical equipment. Worse still, an item may be dropped by a crane so that it pierces the roof or wall of an operating module. To avoid such occurrences the following should be considered.

First, exposed items of electrical equipment located in busy thoroughfares such as walkways, drill-decks and laydown areas should be of steel construction or provided with steel impact protection of some form. No equipment should obstruct walkways. Overhead equipment such as luminaries and cable tray should not project lower than 2 metres (safety helmets add about 100 mm to a person's height).

Secondly, the loci of crane loads should be carefully studied so that areas where the risk of dropped loads is high can be avoided as prospective critical equipment locations. If high-risk areas are unavoidable, then adequate mechanical protection will need to be provided, allowing for the shape and kinetic energy of potential dropped loads.

## Chapter 4

# The Seafloor Structures

## Introduction

In many cases the seafloor has fortuitously, covered and levelled with sediments and consolidated by the action of storm waves. There may be soft, unconsolidated and weak sediments at the seafloor surface.

Rock outcroppings may occur, with highly irregular features. Subsurface strata of sands may be capable of liquefaction under prolonged storms or earthquakes. Unstable deposits at or near the site may give rise to slumping, mudslides, or turbidity currents or may be subject to slow and continued creep. Boulders have been deposited by glacial action on many northern seafloors. Weathering of fractured zones during the sea level lowerings of glacial ages may have produced soft layers between hard rock. Solution cavities may have formed in limestone which is now submerged. Calcareous deposits may have formed on windblown sand as it settled through the water. Recent organic silting or volcanic ash deposits may lie almost undetected on top of competent strata.

Anyone or several of these or other anomalous situations may exist at a specific site. There are two possible solutions; either (1) design the structure to be stable on the actual seafloor soils as they exist or (2) take various steps to improve or modify the seafloor soil properties. The first solution has been the one employed to date in most cases of offshore construction. The second solution is frequently employed for major land structures and is increasingly employed for harbor and coastal (shallow water) structures. The second solution, seafloor preparation, presents some very significant potential advantages for deep water as well. It is being increasingly recognized that there is normally time available in the schedule to do this because of the lead time required for procurement and fabrication of the structure prior to installation.

In some cases, there may also be time after structure installation in which to carry out soil improvement operations; this would be the case, for example, where the structure was installed at the beginning of the good weather season, leaving several months for subsequent soil improvement operations.

It is important always to keep in mind the interactive effects among soil, structure, and the environment. Each acts on and reacts to the others. The environment imposes cyclic loadings on the soil which sometimes leads to physical scour or erosion. The structure imposes forces on the soil, and the soil in turn imposes reactions on the structure. The structure and the waves interact dynamically, as do the soil and the structure, so that as dynamic effects are created in the soils, the soils in turn have a dominant effect on the dynamic behavior of the structure. This process is known as kinematic interaction or soil-structure interaction (SSI).

The adoption and implementation of seafloor preparation measures have been determined largely on the twin criteria of need and practicality. Large over-water bridge piers, for example, have required a high degree of stability and minimal displacements, or tilting. At the same time, such bridge piers have until now been almost exclusively located in water depths less than 100 m and in semiprotected waters. Conversely, with the typical large offshore gravity structure, the critical failure modes have been sliding and rocking. Both of these can be significantly improved by seafloor soil strengthening. Storm-surge barriers are also depend on soil improvement to reduce underflow and to provide support.

With pile-founded structures, lateral stability (the $P/y$ effect) of the piles can be substantially improved, as can the axial capacity.

Seafloor foundation modifications are designed to provide a stable base of adequate strength to support the structure and to resist failure and progressive degradation under both a single extreme event and the repetition of cyclic dynamic loads. The foundation must be graded or levelled as necessary to receive the structure and all obstructions removed. In some cases, protective underwater berms will be placed to protect the structure from ice pressure ridges and ice island fragments or from ship collision. Proper controls must be provided to ensure location and grade and to monitor the performance and adequacy of the measures taken.

Until recently, one of the more difficult tasks was to properly correlate the relative positions of operations on the seafloor with the structure's final location. Electronic navigation and even satellite positioning have not generally been sufficiently simple and accurate in the past to permit accurate relative positions to be determined and repeated.

In shallow water, relative locations can be adequately marked with spar buoys. In deeper water, the articulated buoyant staff buoy provides a permanent marker that is little affected by the waves but strongly affected by the currents. In some cases, it may be possible to use inclinometers with appropriate telemetry in conjunction with these articulated buoys.

Acoustic transponders have now been greatly improved in life and reliability. They can be placed on the seafloor surrounding the site; then their true position can be determined by successive iterations of electronic or satellite position-fixing of the surface control ship. For important structures, where operations will continue for a substantial period of time, enough seafloor acoustic transponders are usually placed to assure adequate redundancy in case of malfunction or destruction of one or more transponders.

Bathymetry can be determined by sonar with due consideration to the relief, contour interval, and motion of the vessel. Corrections must be made for change in water level due to tides, barometric pressures, and storm surges. Corrections must also be made for roll, pitch, and heave of the vessel. On the Oosterschelde Storm Surge Barrier, a remote-controlled bottom-crawling tractor tended by a control vessel was able to map seafloor bathymetry at the site of each pier with an accuracy of +20 mm.

The character of the seafloor can often be determined by video means, using work submarines or an ROV. Side-scan sonar is extremely effective in revealing obstructions and sharp breaks in the surface level. The "profiler," which combines side-scan sonar, automatic compensation, and a directional acoustic beam, is very effective in providing a continuous mapping of the seafloor, especially where there are significant changes.

Existing seafloor soil conditions can be determined from grab samples (for surface classification), by cone penetrometer tests (CPTs), by in-place vane shear tests, by plate-bearing tests, and by borehole sampling. These borehole samples can be obtained by core drilling from a vessel or a work platform, or from the seafloor by jacked sample tubes, or vibratory corers. When deep boreholes are run from drilling vessels, geophysical methods may be used to determine density, resistivity, and permeability.

Experimental work has been carried out with free-fall or explosively driven penetrometers, which send back their changing rates of penetration (deceleration) by telemetry, enabling a determination of relative density at various depths.

Geophysical seismic and near-surface acoustic surveys are very effective in distinguishing anomalies in subsurface geotechnical properties. When correlated with borehole sampling, they serve to portray the area situation much more effectively than linear interpolation between the boreholes by itself.

## Leveling of the Seafloor

Leveling of the seafloor is dependent on having a stable work platform, maintained at a relatively constant grade, from which the drag or screed can be effectively employed. Thus a jack-up rig and a tensioned buoyant platform are especially well suited for such operations.

If the seafloor is generally level but with local ridges and depressions, then dragging of the area with a heavy steel girder can help to smooth out these differences in level. The girder is suspended from a barge by two lines of equal length so that the girder hangs horizontally. As the barge moves across the area, the girder tends to knock down the ridges and fill the valleys between. This method was employed to level the shallow seafloor for the prudhoe Bay Waterflood Facility, a 200-m-long barge-mounted plant.

The difficulties arise with swell acting on the screeding barge, causing the lines suspending the girder to alternately slacken and then become taut. This can lead to the creation of low and high spots rather than their elimination. If the screed girder is suspended by heave compensators on both lines and if the barge or vessel is always headed normal to the swells, then this method can produce satisfactory results during selected periods of low sea states. Alternatively, spars (buoyant columns) can be rigged astern of the vessel to act as crude heave compensators.

Another method that has been proposed and engineered is to use a moored drill ship or semisubmersible with heave compensator from which to suspend the drag. The drag is then pulled along the seafloor by two lines which have been run to preset anchors. A major improvement, now practicable with modern technology, will be to equip the drag head with thrusters controlled from the vessel.

Screeding frames have been developed for use in preparing a level base on which to seat a caisson or an underwater tunnel segment. Some are bottom supported. Such a concept was developed by Christiani and Nielsen for use in leveling the base for breakwater caissons in Cape Town, South Africa, and was later modified for use in seating the caissons for an offshore terminal project off Queensland, Australia, located in the open sea at a depth of 25 m. The concept has been continuously improved for preparing the beds for submerged prefabricated tunnel segments. Horizontal screws have recently been employed for screeding for an outfall sewer pipeline in 100 m of water off San Diego.

Boulders are scattered over much of the floor of the North Sea as well as many other regions. In general, it has been felt that those less than 0.5 m in diameter were sufficiently small that they would be displaced sidewise or pushed down into the underlying clays by the piling or the structure. Boulders larger than this and clusters have been removed.

There are two methods of removal. The most effective one has been to drag the boulders off the site using two tugs and trawl cables and boards, guided by the preset acoustic transponders to the location of the boulder or boulders as previously determined by visual observation from the work submarine. The second method has involved the placement of shaped charges by divers to shatter the boulders; this is obviously limited to those depths and sea conditions in which a diver can effectively work. Thermic lances

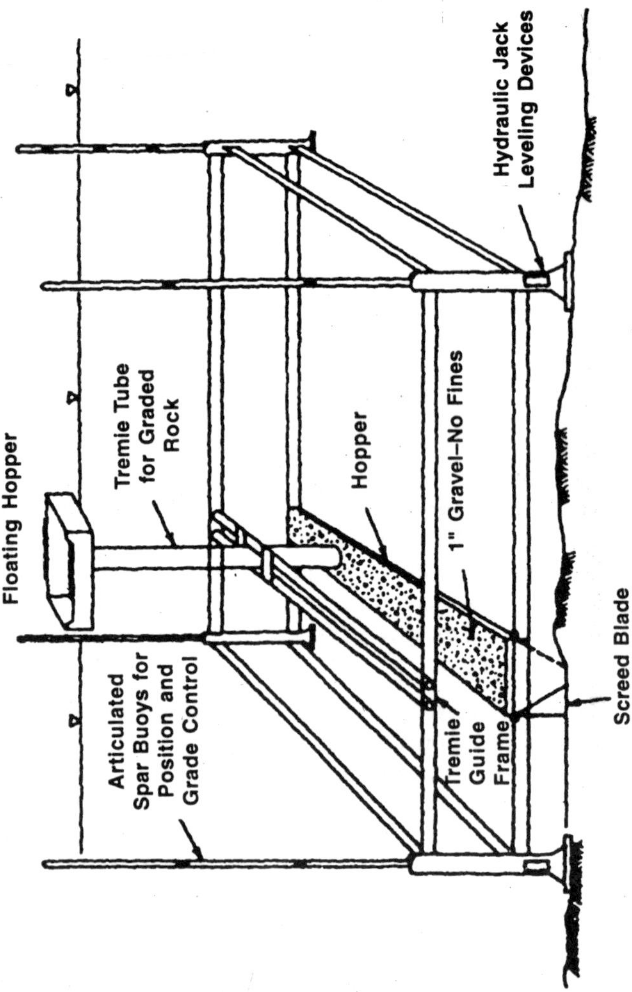


Figure 4.1: Screeding Frame for Underwater Fill–Schematic

can be used to cut large boulders into smaller fragments. Ultra-high-pressure water jets with pressures in the range of 15,000 psi can be similarly employed.

Other obstructions can be removed in similar fashion–that is, by dragging or by individual hooking-on by a diver or ROV to an object previously located visually or by side-scan sonar. In shallow water, *i.e.,* less than 30-m depth, a large grab or clamshell bucket may be used.

For underwater tunnel segments ("tubes"), screed barges have been developed, based on a similar screeding arrangement. However, the supporting platform has been a 100-m-long catamaran semisubmersible barge which lowers heavy concrete block weights down to the seafloor and then pulls down against these to stabilize the barge against waves and the effect of tides. Accuracies of ±50 mm have been achieved. For the Øresund submerged tunnel in Denmark, a standard hydraulic dredge has been fitted with a rotary screed. This is swept back and forth, with its true position determinable by means of electronic positioning and GPS

In the Beaufort Sea, a hydraulic dredge has been used to roughly level preplaced sand embankments at a depth of 10 to 20 m, using a heave compensator to offset the effect of waves. The Dutch similarly used a hydraulic dustpan dredge to level the seafloor for the mattress placement on the Oosterschelde Storm Surge Barrier. In Japan, for screeding the surface of a trench in Tokyo Bay, a horizontal screw (much like a snowplow) was suspended so that it leveled off high spots by moving the ridge material sideways off the site. A similar rotary auger, riding on pre-set girders, was used to level a bed for an outfall off San Diego, California in 100-m water depth. Attempts have been made to screed the seafloor level by diver-manipulated screeds. These have generally been excessively time-consuming and unsatisfactory. A notorious example was the Royal Sovereign Lighthouse in the southern English Channel, where strong tides and storm waves disrupted the work almost as fast as it was done.

For some bridge piers, notably those of the Honshu-Shikoku bridges in Japan, the design has required accurate leveling of bedrock in order to seat a caisson. In such cases, the rock is dredged and then thoroughly cleaned off by jets and airlift. Grinding has then been employed, using large-diameter grinding bits similar to those used on tunnel-boring machines. Horizontally rotating bits tend to "crawl" over the bottom and hence must be rigidly held by a structural frame just above the seafloor. Counter-rotating bits can be used to offset the net lateral forces.

Grinding with wheels rotating in the vertical plane about a horizontal axis is more efficient. These could be extensions of the rock-trenching wheels developed for the English Channel cable crossing and the proposed Straits of Belle Isle crossing between Labrador and Newfoundland. Such a mining tool was used for Pier 7 A of the Bison Seto Bridge pier and for grinding shallow underwater concrete on a lock reconstruction project.

When weak and unsuitable sediments overlie the seafloor foundation soils, they must be removed or displaced or strengthened. In this section, removal and displacement will be discussed.

For large-scale operations in the ocean, one of the most effective tools is the trailer suction dredge. This vessel, usually self-propelled,

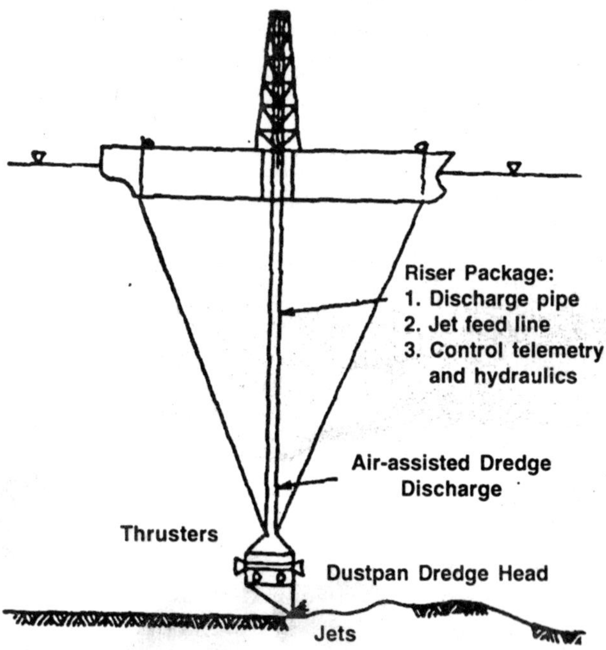


Figure 4.2: Deep-seabed Leveling Device (Proposed)

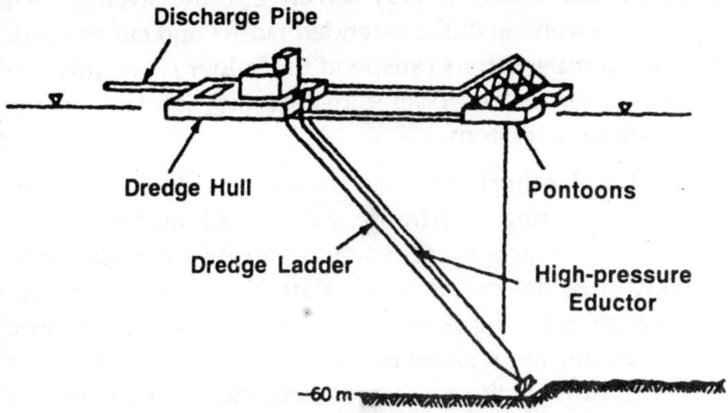


**Figure 4.3: Deep Dredging Cutter Head Dredge Skookum**

uses its speed and momentum, operating through a suspended drag, to excavate the material, which is then sucked up the ladder to the pump and discharged, usually into a hopper for later dumping off site. The trailer suction dredge can take long runs at a site, lower its ladder as it reaches the near edge, and cut a swath across the site in one run.

The drag may be just a steel plate or it may have ripper teeth or jets or even mechanical screw cutters to help cut the soil. This is an extremely economical means of removing seafloor material. It is only limited in depth to the practicable length of a ladder, in the range of 50 to 60 m. It can remove both soft material and partially cemented materials. Its limitation is that it is difficult to control the depth of excavation. Heave compensators have been installed in some cases to keep the drag at a reasonably constant elevation.

The hydraulic cutter head suction dredge is another tool with a long history of successful large-scale application in inland marine construction. This dredge operates most effectively when it is cutting a swath against a face of 1 to 2 m in height, depending on the behavior of the soil. The intent is to have it progressively cave to the cutter and suction but not to bury it. It has been found that cutting the sides of the trench or toe of an embankment "downhill" produces a more stable slope than cutting "up-hill." Dustpan hydraulic dredges have a flat plate that extends beyond the suction. Thus there is minimal disturbance to the soils below grade. In the open sea, the hydraulic cutter head suction dredge is very sensitive to the swell, which aggravates the movement of the extended ladder and cutter head. Use of a heave compensator to suspend the ladder is one positive step. Another is to hinge the ladder from the center of rotation of the hull rather than from its stern.

IHC of The Netherlands has developed an interesting adaptation by mounting a hydraulic cutter head suction dredge arm (or ladder) on a jack-up rig, enabling positive elevation and position control of the cutter head. With this type of dredge, regardless of whether it is supported from a fixed or floating platform, the lateral thrust must be resisted by either the mooring lines or the legs in order to provide the necessary translation and advance of the cutter head. A monstrous walking jack-up was built in The Netherlands to permit progressive advance of the dredge, but proved too cumbersome for efficient operations.

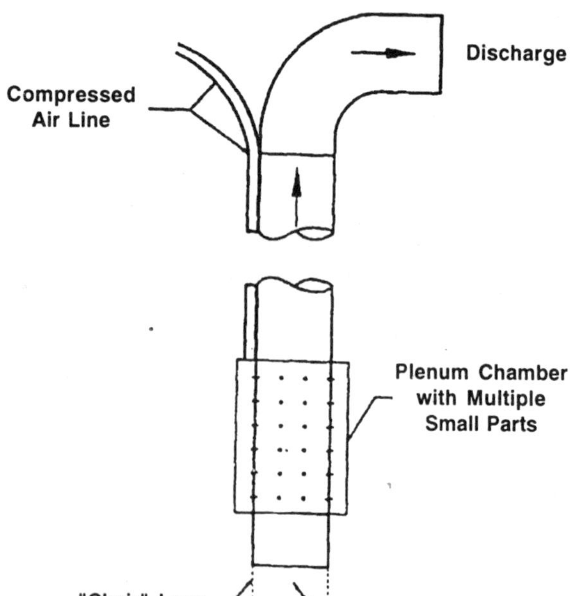


**Figure 4.4: Airlift (Schematic)**

The above schemes are limited in depth to perhaps 50 to 60 m. By use of jet eductor and pumps incorporated in the ladder, the dredge may work to much deeper levels, but it must still be held in position by moorings.

Disposal of hydraulically dredged material in the sea may create a turbidity plume which is environmentally objectionable. A cyclone may be used to separate out the coarser sediments for disposal by direct dumping or barge. Other systems have employed coagulants (thickeners) to precipitate suspended and colloidal materials. Another approach is to discharge down through a suspended pipe so that the discharge is at the seafloor. In the open sea, this discharge should preferably be located below the thermocline to prevent mixing with the surface waters. Deep excavations may also be performed at sea by the use of large clamshell dredges. On the Honshu-Shikoku bridges in Japan, for example, clamshell (grab) dredges have been used to excavate to depths of 50 m or more. These have very large buckets, up to 99 tons. In deep water, the cycle time for such large and heavy buckets is very long. The hoisting time may be reduced by using especially large winches for maximum line pull to reduce the number of parts in the hoisting line and increase the line speed.

The swinging time is again long, due to the inertia of boom and bucket. In some cases, the bucket has been so arranged that it is discharged to a hopper barge moored at the stern of the dredge so that there is no swing of the boom, only a short translation of the bucket along the centerline of the dredge. A 20 $m^3$ orange-peel grab has been used to dig "glory holes" in the seafloor for subsea pump stations. Continuous-bucket-ladder dredges have been used to depths of almost 60 m in the calm seas off Thailand, digging placer sediments for tin.

The airlift, suspended from a barge or vessel, becomes increasingly efficient with increasing depth. The air pressure must be sufficient to overcome the hydrostatic head. It need not discharge at the surface; discharge above the seafloor may be sufficient. The airlift head may be augmented with jets, so arranged as to feed material to the airlift suction. However, this system is effective only over relatively limited areas and for small quantities. The airlift is especially effective when removing material from within enclosures, such as cylinder piles, provided water is continually fed in at the top to maintain the external pressure head. Airlifts are also employed to remove sand and silt from congested zones, *e.g.*, around pile heads in cofferdams. They can be incorporated as air-assists in drill strings.

Other systems for dredging of sediments are the jet eductor, the Marconaflo "Dynajet," and the Pneuma Pump. New submersible pumps equipped with agitators are able to move large quantities of sand at shallow and moderate depths and have proved especially effective in excavation within cofferdams and for cleaning underwater trenches. All are very efficient in removal of material if it is loose and free-flowing. Hence, most such systems include jet systems to break down the soil structure and place the sediments in suspension. In general, these systems are able to increase their production progressively as the water gets deeper.

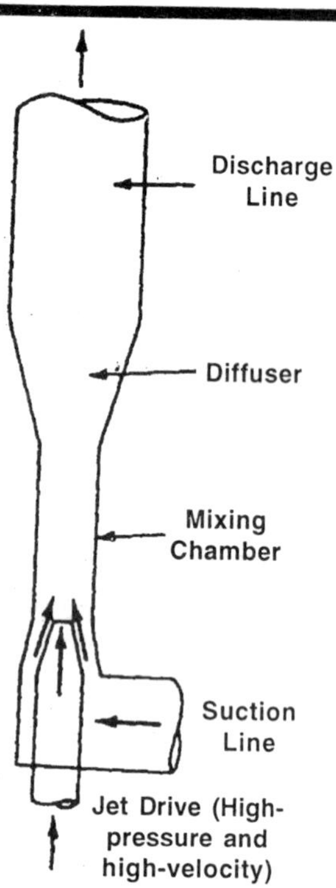


**Figure 4.5: Schematic of Jet Drive (Eductor) Pump**

The jet burial sled, used in pipeline burial operations, is a very efficient method of soil removal, especially in a trench or limited area. It employs the principles of high-pressure jet cutters combined with either airlift or eductor suction and discharge.

To construct "glory holes" in very compact materials for subsurface well templates in areas subject to scouring by the heels of icebergs, an enlarged drill has been used, fitted with a rotating head of 5 m diameter. Reverse circulation methods are employed, with airlift assist, to discharge on the adjacent seafloor. The soil is partially cemented sand and gravel, and extremely dense due to compaction under storm wave action. Successive holes are

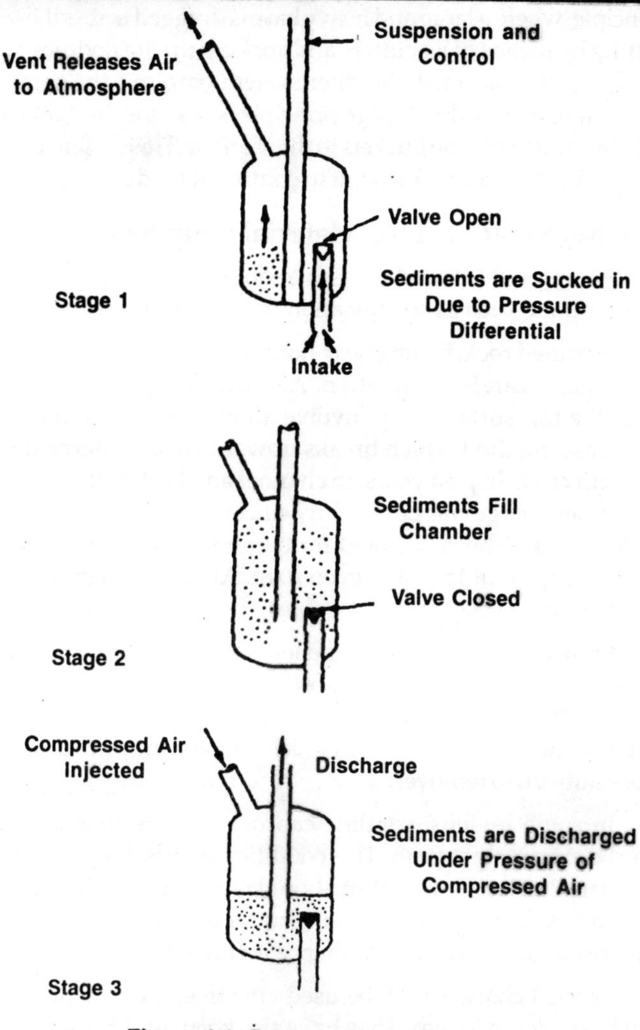


**Figure 4.6: Pneuma Pump Cycle**

overlapped to form a crater approximately 10 m deep, with a bottom dimension of 10 × 20 m. In the dense sands of the North Sea, a heavy orange-peel grab has been effective.

For the deep-sea manganese nodule mining operations, several types of dredge have been developed, which could have applicability to seafloor leveling. One of these employs the suspended drag principle, where a large and heavy base is dragged across the seafloor, cutting by means of jet cutters and sucking up the nodules by airlift or hydraulic transport. Another system employs the continuous-belt dragline or ladder dredge principle (slack-line dredge) whereby a slack line returns the buckets to the seafloor. This system is effective but depth of cutting is difficult to control. It tends to cut trenches.

## The Removal of Hard Material from Seafloor

The removal of hard material requires the consideration of site-specific data, such as stratification fracture and bedding planes.

Stratified rock, having near-horizontal bedding planes, requires particularly careful evaluation. Any dredging system that works from the top surface may involve an excessive amount of effort, whereas a method which breaks upward from underneath can be very effective. In past years, in channel and harbor dredging work, the dipper dredge was particularly efficient. Recently, large hydraulic backhoes have been adapted for barge mounting and are able to work to depths of 15 m and even 20 m. At greater depths, the slack-line dragline bucket should be able to break slabs upward.

Drilled-in explosives are generally inefficient for horizontally stratified and layered rock, as most of the explosive energy dissipates through the cracks. In this case, all that may be accomplished is to fracture the rock into very large slabs, which may make them even more difficult to remove.

In some regions, coraline caprock or limestone layers often overlie extremely soft silt. This condition occurs, for example, in the Arabian-Persian Gulf and offshore Hawaii. Explosives placed on top, such as shaped charges, can shatter the hard layer downward, but are much attenuated and hence inefficient.

Shaped charges may be used effectively to fracture boulders and break down ledges. They have also been used to trench through rock. They must be accurately placed and held in the proper attitude so that the blast creates a cavity in the rock. Divers can place a sandbag over the shaped charge to hold it in proper position. This also improves the effectiveness of the blast. Shaped charge can also be assembled on an expendable steel frame, which is then lowered to the seafloor. Shaped charges have proved especially effective in breaking up surface caprock in order to bury pipelines. Over 300,000

$m^3$ of calcarenite and limestone were broken up with shaped charges in order to bury the submarine pipeline from North Rankin A on the Northwest Shelf of Australia.

Another type of hard material that must sometimes be removed in preparing a foundation is heavy boulder clay. Very large clamshell buckets with teeth can penetrate the clay and engage the boulders with the bucket. One of the disadvantages of the conventional clamshell bucket is that the action of the bucket closing line tends to reduce the effective weight on the teeth. Therefore, hydraulic bucket-closing cylinders have been developed capable of operating at significant depths. These enable the full weight of the bucket to work downward and the full force of the hydraulic cylinders to work sideways, that is, to close the bucket.

Another development of use with clamshell buckets in sandstones and conglomerates is the mounting of heavy vibrators on the bowls of the bucket. When the bucket is on the bottom, with teeth pointed downward, the vibrator is turned on, causing the teeth to penetrate. The proper selection of bucket size and weight, line pull hoisting capacity, and vibrator energy and duration is extremely important. There have been cases, for example, where the vibrator was so effective in penetrating that the bucket could, not be raised; it had anchored itself into the rock.

Cutter head suction dredges have been developed with tremendous power on the cutters so that they can cut soft to moderate strength rock. They have been used to deepen the Suez Canal, removing sandstone, and to mine rock salt in the Arabian Gulf. However, the action of the typical cutter, rotating around the axis of the ladder, is a grinding action that may involve excessive expenditure of energy. Another new development is the cutter which rotates around a horizontal axis (the wheel cutter head), designed to break the material upward, where it can be picked up by the suction.

In many cases the material may be pre-broken to facilitate removal. For example, in boulder clay, high-pressure water jets may be used to erode the clay binder, enabling dredges to work more effectively. Surface explosives may break the cementitious bonding of conglomerate formations. Drilling and blasting with light charges will greatly increase the productivity of hydraulic dredges in weakly cemented or overconsolidated materials.

For an intake pipeline in a deep mountain glacial lake in Alaska, the overconsolidated silt was drilled using a jet pipe and high-pressure water. These holes were on a spacing of 3 m. They were then loaded and blasted, using delays to cause the soil to break toward an open face. A hydraulic cutter head suction dredge was then able to remove the soil efficiently, provided it was activated within 1 or 2 days after the blasting of a particular cut.

A recent development for quarrying operations on land, which may someday have application at great depth, is the use of hydraulic fracturing techniques. Short bursts of extremely high-pressure water (up to 15.000 psi) are used to propagate fractures in the rock.

The use of underwater chisels is a method of rock breaking that avoids the use of explosives. In relatively shallow water (15 to 20 m) the chisel may be a heavy piece of shafting, extending up above water. It may be repeatedly raised and lowered to fracture a hard layer; this rather crude but effective process has recently been employed on a large scale in the Arabian Gulf port projects. In some cases, a high-pressure water jet has been incorporated into the chisel to wash away loose and broken material.

A more-controlled operation is to use an impact or vibratory hammer on top of the chisel, thus driving it into the rock. After penetrating a meter or two, the long chisel is pulled sidewise at the top, breaking off a piece of rock, just like a gigantic clay spade. On some large projects in the Arabian Gulf and the Mediterranean, a bank of such chisels is assembled along the side of the rock-breaker barge to methodically break up a hard rock layer for subsequent removal by a hydraulic dredge.

Rock-breaking chisels, driven by hydraulic or vibratory hammers, can also be operated underwater. Their location must be carefully controlled. They use their weight plus impact or vibration to penetrate. Incorporation of a high-pressure jet may help to dislodge the broken rock and prevent "self-anchoring."

Large hydraulic backhoes, mounted on a barge equipped with spuds, are very effective in digging soft and layered rock to a depth of 20 m.

Drilled-in explosive fracturing has a long history in underwater rock dredging. The hole must be cased from above water down into firm material, usually to top of rock. The casing is either driven in through the overburden or drilled in. This latter is known as the "OD Method." After the hole is drilled and cleaned, the waterproofed charges of powder are lowered down with either waterproof leads or primacord attached. Sand packing is placed on top of the powder (stemming), and the leads or primacord are led all the way out of the casing at the top, with a float attached. The casing is now raised, and the leads are picked up and connected on the barge. After a series of holes have been so charged and connected, the barge pulls away 60 to 100 m and the round is fired.

The effectiveness in shattering rock is greatly increased by the amount of overburden. Also, the presence of overburden makes it easy to seat and seal the casing, whereas the absence of overburden makes it extremely difficult. Hence, all or some of the overburden should be left in place during the drilling and shooting operation.

Typically, holes are drilled on a 2- to 3-m spacing. The spacing of holes and overdepth (below design final grade) of drilling must be carefully determined to obtain optimum results. From the dredging point of view, aimed at facilitating rock removal, it is generally best to drill below grade a depth somewhat more than half the spacing of the holes to ensure against high points (pinnacles) being left above grade. For example, a crude and conservative rule of thumb is "half the spacing plus half a meter." The reason for this is that it is almost prohibitively expensive and difficult to attempt secondary shooting on high pinnacle remnants. Staggered rows of holes appear to give better fragmentation than a rectangular grid.

For the same reason it is considered good practice to use a conservatively high powder factor (*e.g.*, 1.2 to 1.8 kg/m$^3$) and a relatively fast powder (*e.g.*, 60 per cent). However, such a procedure may cause excessive fracturing of the rock below grade, which may or may not be acceptable from the foundation point of view. If not acceptable, the rock may require subsequent mechanical grinding or pressure grouting, this latter being usually done after the structure is installed. On the Tsing Ma Bridge in Hong Kong, plastic tubes were embedded in the concrete footing block. Later, they were used to drill and grout the fractured rock beneath.

If it is necessary to minimize the subsurface fracturing, then holes should be drilled on a closer spacing and to a correspondingly lesser depth below grade. Smaller charges should be employed, *e.g.*, less than 1 kg/m$^3$.

Current safety regulations prohibit drilling within 15 m of a loaded hole, unless a waiver is granted. Such a waiver has sometimes been granted if a template is employed to control the spacing and verticality of the casing. However, the rules can be complied with by using a large template, to enable a full line and one or two rows of casings to be left in place until all are drilled, then load them.

Blasting of rock is always most effective if there is a face toward which the rock can break. Therefore, in some cases it may be desirable to drill and partially excavate a trench, then progressively drill, blast, and excavate, so that there is always an open face. If there is no open face, then there is a tendency for the rock to fracture, raise, and settle back in the same compact mass. This can be particularly adverse if the powder factor is low and a slow charge is used; the effect will be to fracture large pieces without displacement. Subsequent drilling and shooting may then be very difficult, since the blast will dissipate along the fracture zones without breaking new rock.

When adjoining structures must be protected, line-drilling and cushion-blasting techniques are employed. Presplitting along a boundary row prevents extension of the fractures. Use of delays can ensure that the blasted material moves in the desired direction. Air bubbling reduces the water shock effects. By placing a double row of air bubblers at 3 m spacing, and 3 m outside the structure to be protected, the peak water pressure can be reduced by a factor of 10.

If structures have water on both sides, *e.g.*, a pipeline is full of water, then the damage is greatly reduced, particle velocity can be limited, *e.g.* to 12 mm/s, and water overpressure should also be limited, *e.g.* to 0.5 N/mm$^2$ at the face of the structure. Stemming the top of the hole with a depth in the top rock equal to half the hole spacing is effective in reducing damage and turbidity. Test blasts can be utilized to establish an effective value for powder factor.

The blasting of solid rock produces fractured material having typically 40 to 50 per cent greater volume, thus raising the level of the seafloor in the blasted area. If the material is to be removed by bucket dredging, the bulk quantity will be increased accordingly. This also applies to the rock, which has been blasted below grade. Thus extensive areas of shallow rock removal may involve a 100 per cent or more increase of the dredged quantity as compared to the neat solid rock volume.

For deep offshore work, either rotary or percussion drills are employed. Rotary drills are best for deep drilling in competent rock, whereas pneumatic drills are most useful in irregular, erratic material and shallow drilling depths. Pneumatic down-the-hole hammer drills are proving effective and economical in underwater shaft construction and the construction of rock sockets for piling. The use of a rotary drill, supported by a large jack-up, to level the rock bottom for the anchorage pier of the Bison Seto Bridge between Honshu and Shikoku.

Intake shafts have been constructed in the deep water of existing lakes, while outfall risers have been constructed in the subsurface rock of coastal waters. These may be drilled with full-face rotary drills up to several meters in diameter. An alternative means is to construct a seafloor template for guidance and then drill numerous 300 to 500 mm diameter holes by a down-the-hole drill. These are very closely spaced, to create a "Swiss cheese effect." The interstices are then broken down by a spud chisel. In all cases, the hole has been cased and reverse circulation methods employed.

Drilling and blasting can also be carried out by divers and/or submersible work vehicles. This has been, effective only for relatively small, isolated features and shallow depths of work, such as isolated boulders on the seafloor. However, divers are generally not able to carry out major operations over an extended period of time. Their effectiveness is also limited by their inherent buoyancy. Underwater tracked vehicles have been experimentally developed to carry out the drilling from the seafloor. For any major project, working from the surface is currently the most efficient and economical.

Considerable work has been carried out with acoustic underwater blasting devices to eliminate the need of bringing leads or primacord up through the water for collection on the barge. These devices appear to be reasonably reliable, provided they do not become silted over. They were effectively used on the rock excavation for the piers of some of the Honshu–Shikoku bridges in Japan. Subsequent to blasting and excavating rock, it is generally necessary to clean up the foundation by removal of silt, sand, and small rock fragments. This is best done by a straight suction dredge or airlift, aided by high-pressure jets.

Turbidity has recently become a major concern for underwater blasting. Tests in Japan have shown rapid sedimentation of coarser silts, but progressively lower rates for fine silt fractions and clay particles. For these, silt curtains are employed. The most effective curtains have been supported on a large structural frame, through which the bucket is raised and held above to drain the water before swinging. In discharging fine material underwater, it should be led down a tremie to discharge just above the bottom.

## The Techniques of Underwater Fills

Underwater fills are often placed by discharge from a hydraulic or hopper dredge or by dropping from a bucket to fall through the water.

When low-relative-density materials are placed through water, they tend to disperse laterally and to fall through the water at differential speeds. The result is to segregate in layers of different size. In addition, the in-place density of such material is heavily dependent on the permeability and relative gradation of the particles. Relative densities of cohesionless materials (sands and gravels) placed through a substantial depth of water may vary from 40 to 60 per cent, with 50 per cent being most common. Lateral spreading is dependent on specific gravity, gradation, particle shape, depth, and currents, but in general slopes are very flat. As fine sands impact, they temporarily liquefy, allowing them to flow locally as a dense fluid. Silt or mica content is very critical: lenses formed during deposition can lead to slope failures at a later stage. If these fills are later cut with a bucket or cutter head, they will stand at a temporarily steeper slope but will be very sensitive to shock and vibration.

Air content at the time of placement has a very significant effect on segregation, spreading, and density of in-place underwater fills. The air bubbles attach to fine particles and give them added buoyancy. The tendency for such segregation can be reduced by thorough saturation of such materials prior to placement.

Sand may be discharged down a tremie pipe whose end is fitted with a special device to force the sand and water to separate and hence enable a steeper slope to be attained. Among the special devices employed are screens of "fabric" mesh, whose openings allow the water to flow out freely while tending to restrict sand passage, and a wide bell-mouth fitting which reduces the exit velocity. At the Tarsiut Island in the Canadian Beaufort Sea, for example, use of such a device resulted in a side slope of 5.5 : 1 as compared to slopes of 10 : 1 and even 15 : 1 when the sand was discharged at the surface.

Underwater fills of granular material such as crushed rock and gravel can be used to provide a reasonably level and uniform support for structures at a practicable and economical level. For example, they can be placed over irregularities and outcrops or used to fill back depressions from which unsuitable materials such as mud have been removed. In deep water they can be used to raise the base of the structure to a more favorable elevation from a standpoint of economy while still staying well below the elevation at which the design wave will have destructive effect. Such fills can also provide a foundation of known static and dynamic properties from the points of view of stiffness, pore pressure buildup, and resistance to liquefaction.

Protective islands are built surrounding bridge piers, to prevent ship impact. Fill is placed around and over submerged tunnel segments and outfall sewers. Embankments are constructed to channel rivers, for approaches to piers, and behind wharves as the storage yard for containers as well as a dike on which to construct the wharf. Fills are also placed underwater to form the core of a breakwater.

Underwater fills of crushed rock or gravel may be used to blanket an area to contain unstable sands and allow pore pressure relief without sand dispersion. They may also be used to laterally confine unstable materials such as clays, acting as counter balancing surcharges external to the structure, thus preventing local shear failures.

Underwater rock blankets may be used to cover over irregular rock outcrops to permit a structure to be founded with uniform bearing. A clay blanket may be used to blanket contaminated soils. Underwater rock dikes, placed prior to or during dredging, may be used to stabilize side slopes against shear and erosive failures. In such cases the underwater rock dike migrates downslope and into the sand as the dredging takes place, serving to give steeper and more stable slopes. These rock dikes are known as "falling aprons." They are most effective when no filter is used beneath; thus the sand progressively migrates through the rock, allowing it to fall on a regular pattern. Clay dikes, using stiff glacial clays, have also been used to retain underwater sand fills.

The materials for underwater fills have to be selected with regard to their suitability for the needed objectives, their density and size gradation, and their ability to be placed at the depths and locations desired. Obviously, availability and cost are also factors.

If an underwater fill is to contain sands and prevent sand dispersion through the fill, then the material should be graded as a filter. In practice, this is extremely difficult to accomplish. One approach is to select a well-graded material similar to a concrete mix that will act over the complete range both as a filter and as a stabilization and erosion protection; some fines will be lost, but the remainder will stabilize. Another solution is first to cover the area with filter fabric mats with articulated concrete blocks attached. Then rock can be dumped over the mat. Mats can also be held in place by sandbags or steel pins set by divers. In very deep water, filter fabric mats may be preattached to the structure prior to seating it or may be attached to steel or concrete frames or panels.

One of the best methods of placing underwater rock fills in the ocean is by bottom-dump or side-dump barge. By pre-saturating the material prior to dumping, segregation is minimized. The mass of the rock hangs together as it falls through the water, thus attaining the terminal speed of the mass, which is considerably faster than that of the individual particles. The impact of the mass on previously placed material helps to consolidate it.

Tests in The Netherlands by ACZ Marine Contractors have shown that a mixture of stones with a maximum diameter of 0.2 m (200 mm) will reach a maximum terminal velocity of 2.0 to 2.5 m/s. However, when dumped as a mass from a split hopper or similar barge, the entire mass hangs together developing a fall velocity about twice that of the individual stones, that is, 4 to 5 m/s. These terminal velocities are reached within a relatively short fall distance (*e.g.*, 5 to 10 m) regardless of the initial velocity of the rock falling through air or through a pipe.

Other methods that have been employed involve placement by clamshell bucket or skip lowered to the bottom before opening (a very tedious process) and placement through tremie pipes. These latter are usually suspended from a barge or from a pontoon-supported hopper, laterally restrained by lines to the barge. The pipe must have a large enough diameter to avoid any possibility of plugging: a value of three to five times the diameter of the largest rock pieces is often used. Gravel is less likely to plug than crushed rock. Elongated particles are unsatisfactory. Another method of placing rock at depths is through an inclined chute such as the modified ladder of a trailer suction hopper dredge. The discharge end is suspended from the vessel by a heave compensator and can be directed to the proper location.

A number of rock-placing vessels have been developed by Dutch dredging contractors which are equipped to place rock either by direct dumping or through a controlled tremie pipe.

The selection of proper particle size for long-term stability requires consideration of bottom current velocities due to combined tidal and general currents and storm-wave-induced currents. The effects of the structure itself in generating vertical and eddy currents must be considered. During a storm, the pore pressures in the soil will fluctuate and may make it easier for fill particles to be temporarily placed in suspension.

Another consideration, of course, is the packing factor, which is determined by particle shape, gradation, and degree of consolidation of the fill. Relative density is extremely important, since we are dealing with the submerged density. Use of a rock consisting of iron ore mineral compounds and having a specific gravity of 3.5 or higher is much more effective and stable than the typical silica rock with a specific gravity of 2.6, since stability varies approximately as the cube of the underwater net density. Larger sizes are also more stable and can serve to lock together a fill of varying gradation.

When fills are placed around a structure–for example, when sand is discharged down a tube as backfill for a pipeline–the fill is temporarily a heavy fluid and has the flow and displacement properties of a fluid. Thus it can run under the pipeline and lift it up or displace it sideways, just as if it were a fluid having a specific gravity of 1.5. A number of major pipeline installations have been seriously dislocated or even ruptured during backfill in this manner. Of course, the material quickly returns to a solid once the excess pore pressure has dissipated, which usually occurs in a few seconds, but the damage may already have been done. This same problem but with even greater consequences can occur when underfilling and backfilling prefabricated submerged tunnel segments.

The proper use of underwater fills would appear to be a major opportunity in the extension of structures and consolidation to deep water, to less competent soils, and to more exposed locations. Properly employed, the fills can enable construction to be carried out at a more favorable elevation with materials of known and controllable properties.

Underwater rock dikes placed around a structure can be used to prevent collision from ships, to cause icebergs or deep pressure ridges to ground, and to divert mudslides and turbidity currents. They can also, as mentioned earlier, confine soft soils such as clays, in order to lengthen the shear path and to resist local shear failure due to high bearing pressures. Particularly attractive is the fact that in many cases the underwater rock fills can generally be placed

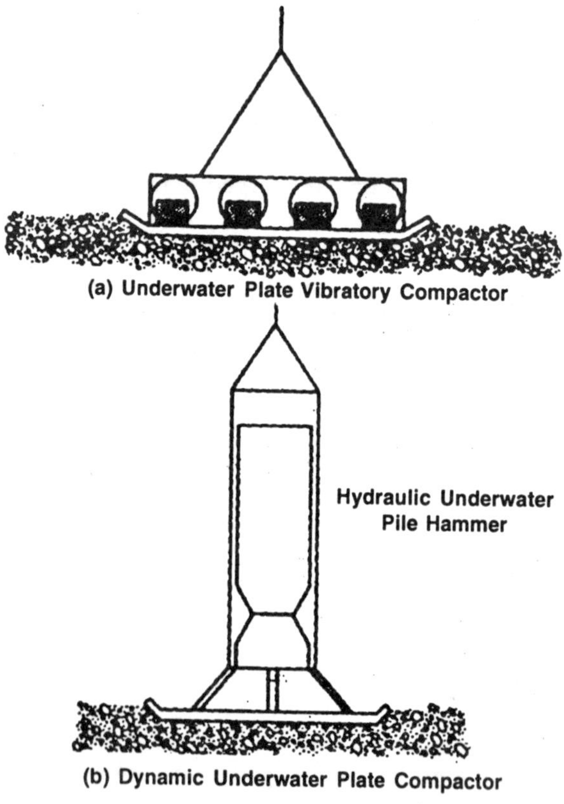


Figure 4.7: Compaction Equipment for Underwater Rock

during the period in which the structure itself is being fabricated; thus, the work is not on the critical path.

In-place sands may be so loosely consolidated that they will be subject to liquefaction under prolonged storm or earthquake. Densification procedures are therefore employed to raise their relative density above the critical valve (about 70 per cent R.D.). Fills such as those described in the preceding section may also require consolidation in order to reduce settlements and to ensure stability. There are a number of methods and techniques by which soils and fills may be consolidated. One of these, which applies to granular, cohesionless materials, is vibratory consolidation. A large mandrel–up to 1 m in diameter, for example–is inserted into the material by either jetting or vibratory driving. The mandrel has horizontally oriented vibrators mounted in its tip, which are now activated, imparting high-frequency energy into the adjacent soil. This causes the sand particles to reorient themselves. Pore pressures are increased and then relieved by drainage through the adjacent fill.

Several brands of internal vibratory compactors are currently manufactured, some of which are able to work entirely underwater. These are able to penetrate and consolidate material ranging from 75 mm diameter down to fine sands.

It is essential that the pore water be able to drain. Layers of silt or clay will prevent drainage and seriously limit the efficacy of the vibration. If these blanketing layers exist, vertical drain paths such as gravel drains must be provided. These can be installed as drilled wells or even jetted into place. There must also be a horizontal escape path for the water that is expelled; this is usually a sand or sand and gravel layer preplaced on the seafloor, under the base of the structure. Alternatively, drainage may be provided into the structure, from where it is pumped out.

Internal vibration does not compact the near-surface layer. For this layer, a vibratory plate compactor must be placed on the surface. A large vibratory plate compactor was employed on the Oosterschelde Storm Surge Barrier where it successfully compacted rock of 350 mm maximum size in layers up to 4 m thick. A similar plate vibratory compactor was used to densify the soils on which the Great Belt approach piers and main pylons were founded.

Other methods of surface compaction utilize adaptations from above-water landfill practice; these include a roller compactor and a remote-controlled underwater bulldozer. These, however, have so far been limited to relatively shallow water. Another crude but effective tool consists of a long shaft or pipe with a plate on the bottom. A pile hammer or vibration hammer is attached enabling effective control. Another method, adaptable to a wide range of material sizes and gradations, is dynamic compaction (the "Menard system") which involves the repeated raising and dropping of a heavy weight. Depending on its mass, density, and distance of fall, this dynamic compaction can effectively consolidate up to as much as 10 m of underwater soils or fill. This has been used successfully on both sand and rock underwater embankments. However, the effect of the shock on the adjoining sediments must be considered, so as not to cause large-scale liquefaction.

Air guns, as used in geophysical seismic surveys, may be used to densify loose sands; some experimental work has been carried out. It is important to recognize that this shock causes high pore pressures to build up instantly. These are then relieved by drainage through permeable materials or through shear fractures in relatively impermeable materials such as silts. This latter is, of course, normally undesirable and may result in slope failures.

Similar "shock" or dynamic compaction can be achieved by use of explosives. These are installed by jetting a casing into the fill to about two thirds of the depth of the stratum to be densified. The size of charge is limited so that craters will not form. Typical spacings of holes are 3 to 8 m. Delays should be used to separate the times of successive impact.

## The Scour Protection

The most common form of scour protection is by the placement of rock of a size suitable to withstand the currents without dislodgment. This may vary in size from gravel on a seafloor at a depth of 20 to 30 m to 10-ton rock in the surf zone. The stability of rock varies as the cube of the buoyant density, that is, specific gravity$^{-1}$. Hence, trap rock with a unit weight of 190 lb/ft$^3$ was specified for the breakwaters of the Atlantic Generating Station planned to be built off the coast of New Jersey in lieu of somewhat less expensive silica rock at 165 lb/ft$^3$. A similarly sized piece of trap rock has a 50 per cent greater stability factor than that of silica rock. Similarly, iron ore has been considered for protection of a river bottom against scour, where the size of rock was limited by other considerations.

In general, the larger the individual pieces, the greater their stability. However, interlocking of pieces is also very important, with blasted polygonal rock being much more stable as a mass than similarly sized cobbles and boulders. Hence, chinking of the crevices and even filling them with concrete can add to the stability as long as there is sufficient permeability (porosity) to allow excess pore water pressures to dissipate.

Wave impact creates hydraulic ram effects that temporarily raise the internal pore pressure. A breaking storm wave can create a hydraulic ram effect that can (and has) hurled a 100-ton block over the breakwater! Breakwater armor typically fails outward, at least initially, due to these effects. Significant wave forces can extend to depths as great as 100 m, where they can create internal pore pressures of 3 to 14 tons/m$^2$ (30 to 40 $k_a$).

Under wave action, the sand immediately under rock fragments liquefies. The sand particles then migrate through the rock and are washed away, allowing the rock to work its way down into the sand. To prevent this, filter courses or a mat of filter fabric are placed, graded so that the sand will not work its way through the rock.

The Shore Protection Manual of the U.S. Army Corps of Engineers gives specific guidelines for the sizing and gradation of rock to serve as filters. Placement of several layers of rock of different gradations is difficult enough in the calm water of harbors but much more so in the open sea. It becomes necessary to increase the thickness of each layer to compensate for the tolerances in placing. The most practicable way, albeit inefficient in the use of material, is to mix all the materials of different sizes to grade from fine to coarse and then place as one combined layer. A mix of aggregate suitable for concrete (but without the cement) has been frequently used to stabilize the bottom of a shallow cofferdam.

Filter fabrics have today been widely adopted. These fabrics have specified sizes of pores, which will allow water to bleed through but not sand. To give adequate strength to such fabrics to accommodate differential movements, wave and current forces, and the strains induced during installation, the finer fabrics should be backed up by heavier-mesh polypropylene. The heaviest such material even has embedded stainless steel strands. Fabrics of the two types may be sewn together and laid as a unit.

Most of the filter fabrics are lighter than water and hence hard to place underwater. Concrete "dobe" blocks may be attached with

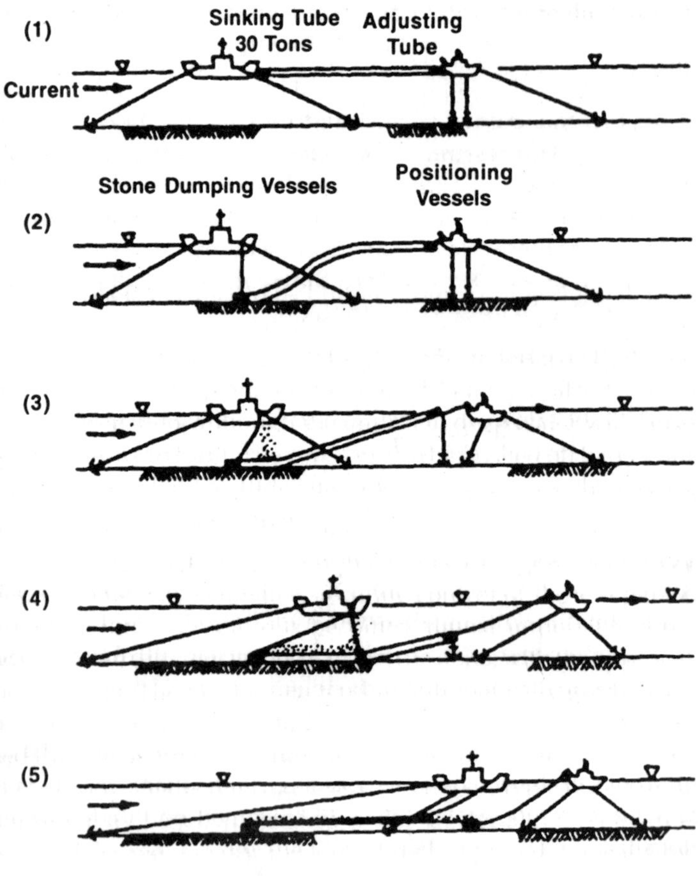


**Figure 4.8: Procedure for Sinking Protective Mattress with Stone Cover**

**(1) Position equipment; (2) Sink "sinking tube;" (3) Start stone dumping. Shift positioning vessel; (4) Continue stone dumping. Stretch mattress; (5) Sinking completed.**

stainless steel staples. The Dutch automated the manufacture of such articulated mats for the Oosterschelde Storm Surge Barrier. In Canada, special filter fabrics are manufactured which are denser than water and hence more easily installed. These mats may be laid as rolls, unreeled off a large drum, and spread onto the seafloor. Alternatively, they may be assembled on steel pipe frames, say 20 × 20 m in plan, and set on the seafloor. Mat segments such as these were installed around the offshore terminal caissons at Hay Point, Queensland, Australia, and have successfully withstood the scouring attack generated by cyclones as well as high tidal currents. To protect the underwater embankments adjoining the Jamuna River Bridge, Bangladesh, fascine mattresses of bamboo, filter fabric and stone were placed on the slopes.

Other ingenious forms of scour protection have been developed. Holes left in an external wall of the structure near the seafloor may develop counter-currents that tend to cause deposition of sand rather than erosion. Steel pipe frames, cantilevered out from the structure just above the seafloor and hung with multiple strips of plastic, can act as artificial seaweed, slowing the local currents, causing deposition.

Much experimentation has been carried out on the laying of a mat of concrete or asphalt over the seafloor in order to stabilize it. The Dutch have applied asphalt-sand and asphalt-stone mixes on many of their coastal dikes. These are usually designed to have both flexibility to accommodate local movements and porosity to relieve excess pore pressures. German and Japanese engineers have developed special concrete mixes and admixtures such as silica fume and anti-washout to enable a permeable concrete mat to flow out over the seafloor without segregation.

Where rock riprap of the required size is unavailable or excessively costly, concrete armor units have been used. There are at least 40 different shapes which have been developed, of which the Tetrapod, Tribar, Stabit, and Dolos are among the best known. The Dolos has the best hydraulic characteristics and least material quantities but is weak structurally in the larger sizes. Breakage of the 60-ton Dolos units has been assessed as the principal cause of the failure of the Sines offshore breakwater in Portugal. The Waterways Experiment Station has developed a new breakwater armor unit which appears to optimize hydraulic and structural performance.

Reinforcement has been used for some U.S. armor units, and experimental application of steel fibers in the concrete mix has been tried on the Crescent City breakwater, in California, with apparent good results.

For offshore sand and gravel islands in the Arctic, filter fabric has been laid over the slopes near the waterline and then 2- and 4-$m^3$ sandbags of polypropylene laid in one or two layers. The polypropylene is pigmented to prevent ultraviolet (UV) disintegration. Significant damage has occurred, however, due to sea ice attack on the sandbags, especially while they are still frozen. Articulated concrete blocks over heavy filter fabric are believed to give more permanent protection. These have been used on the Endicott Production Island in the Alaskan Beaufort Sea with generally good performance.

Another form of slope and scour protection is formed by tubular sacks of plastic. These are laid out on the surface and pumped full of grout. They are joined by integral webs of plastic, having small holes to allow excess pore pressure to escape. For the Oosterschelde Storm Surge Barrier, mattresses were fabricated, having three layers sewn together. These layers were filled with sand and stones to form a graded filter. The mattress was then unrolled over the seafloor. A similar layered mat is planned for the bottom of the excavated trench for the Olmsted Dam on the Ohio River. This mat will include a submattress filled with bentonite to allow the dam to translate independently of the soil under earthquake. From the constructor's point of view, the use of the fabric mat with articulated concrete blocks attached seems to be the easiest and most positive method of providing scour protection for offshore structures.

Permanent scour protection is generally the province of the designer. However, it will often be found necessary for the constructor to place temporary scour protection. An example is the landing of a caisson on a sand seafloor where the local currents are quite high; these could then accelerate under the structure's base and cause serious scour just prior to touchdown.

On the box caisson pier for the Øresund Bridge between Denmark and Sweden, storm-induced currents and waves undermined the pier shortly after it was installed. It required an extensive grouting program to fill under the base. A filter course of stone was then placed around the pier and covered with heavy riprap. Many similar cases of undermining of bridge piers have occurred on rivers during floods, since the scour increases exponentially with the velocity of the current.

During construction of a bridge across the lower Colorado River, the increased blockage of the river by the cofferdams for the piers caused an increase in velocity, even though there was no flood. Since the sand at the site was very loose, it scoured away, in some places all the way to bedrock. During construction of a bridge across the Skagit River in Washington, the blockage was caused not only by the cofferdams but also by the contractor's fleet of barges moored alongside.

Now let us learn few things about the consolidation and strengthening of weak soils.

Sand piles or stone columns are installed by drilling or driving into clays and silts in order to strengthen them by increasing both bearing and shear resistance. Such piles are usually installed by a driven pipe mandrel into which the granular material is fed. This material is then forced out with air pressure as the mandrel is withdrawn. Since only limited depth of penetration is required–that is, only the weak soils are involved–the process is very rapid. Typically, 1-m-diameter "piles" are placed on 3-m centers. These are often referred to as "sand drains," "sand piles," and "stone columns." A similar process uses a jet to assist in sinking the pipe mandrel, then feeds gravel into the column as it is withdrawn, while vibrating intensely.

Vibratory compaction, as described in the previous section, is extremely effective in consolidating loose sand deposits. Perhaps the most extensive use to date was on the Oosterschelde Storm Surge Barrier, where four such compactors which were mounted on a barge compacted the loose sands to a depth of 50 m below sea level. For the piers of the Great Belt Bridges, heavy vibratory compactors were used to compact the stone-fill base, working in 1-m layers to achieve 80 to 90 per cent relative density.

An extremely effective means of consolidating weak soils is by surcharging, with subsequent removal or redistribution. An underwater fill of sand or rock can be placed by bottom-dump barge and allowed to exert its excess pressure on the soil for a period of 6 months to year or more. The consolidation will be even more effective if the pore water has an opportunity to drain through natural or artificial permeable drains.

Silts and clays may be consolidated by means of drainage using vertical sand drains or wicks. This method is especially applicable if the soils are anisotropic with good natural horizontal permeable stratification, as exists in many locations where the sediments have been deposited in successive layers. This system, much used in land operations, can also be used underwater if accompanied by a surcharge.

Sand drains can be drilled in or jetted. Wicks are installed by jetting or by attachment to a dropped shaft which penetrates dynamically. Of course, after the structure is in place, drainage wells can be drilled in and even pumped to remove excess water during consolidation. The water that is driven out of the soil moves upward through the drains and must then have a means of escaping laterally. Thus a blanket of coarse sand and gravel should first be installed over the seafloor area. This blanket can also be a surcharge.

The vacuum process has been used to accelerate the consolidation by drainage of above-water fills. After the wick drains have been installed, an impervious plastic membrane is laid over the area and sealed at the edge by fill. A vacuum is then drawn and maintained. This process has the advantage that it can be used over very weak soils for which the more conventional surcharge might cause shear failures. It was employed on the new terminal at Neva Shiva in Bombay.

With or without artificial drainage the surcharge effect of a fill over a period of 6 to 12 months may serve to stabilize silts and possibly even clays to an acceptable strength to receive and support the structure. Then just prior to the installation of the structure, the surcharge fill can be spread laterally to a peripheral location where it will lengthen the shear path and act as a counterbalancing force against bearing failure.

Cementation of underwater granular soils can be carried out by injection of cementitious material, following land-based grouting procedures. The cementing pressures must be regulated to displace the pore water yet not cause "fracturing of the formation" through channelization. Cementitious particles must be small enough (fine-enough grind) to penetrate the interstices. Addition of a wetting agent to the grout which reduces the viscosity and colloidal mixing facilitates permeation.

**Table 4.1: Seafloor Modification and Preparation**

| *Task* | *Methods* |
|---|---|
| (A) Survey, investigation, and controls | GPS, DGPS, lasers, electronic navigation systems, spar buoys, acoustic transponders, coring and sampling, grab samples, sparker survey, side-scan sonar, acoustic imaging, foundation penetrometers, video submersible and diver inspection. |
| (B) Platform | Crane barge, drill ship, barge, semi-submersible, jack-up, TLP, heave compensators, guyed tower. |
| (C) Seafloor obstruction removal | Drag off with trawlers, shaped charges, ROVs with manipulators, underwater burning, thermic lancers. |
| (D) Dredging removal of sediments | Trailer suction hopper dredge, cutter head hydraulic dredge, clamshell (grab dredge), dragline, continuous bucket ladder dredge, slack-line bucket dredge, plow, jetting, pipeline burial sled, deep-sea mining drag excavator, airlift, eductors, remote-controlled seafloor dredge. |
| (E) Dredging removal of hard material and rock | Hydraulic backhoes, dipper dredges, power-activated clamshell buckets, plows, shaped charges, blasting in drilled holes (O.D. method), chisels, hydraulic and pneumatic rock breakers, driven spuds, cutter head dredges, high-prescure (15,000 psi) jets, down-the-hole drills. |
| (F) Placement of underwater fills | Dikes of rock or clay bunds to contain sand, controlled underwater deposition, dump en masse from hopper barges, tremie, bucket, skip, chute, or ladder. |
| (G) Densification, consolidation, and strengthening of fills | Deep vibration, surface vibration, dynamic compaction with dropped weights, explosives, or air gun, deposition in mass, pre-saturation, cement grouting, chemical grouting, closely spaced driven piles. |
| (H) Consolidation and strengthening of weak soils | Sand piles, vibration, freezing, surcharging with membrane and drainage, surcharging with structure and ballast, wick and sand drains, drainage wells, peripheral surcharging, cement injection, chemical grouting, lime injection, deep cement mixing, electro-osmosis, jet grouting, stone columns. |

*Contd...*

**Table 4.1–Contd...**

| Task | Methods |
|---|---|
| (I) Prevention of liquefaction | Densification as per H (above), drainage wells, peripheral apron of graded rock, stone columns, wick drains. |
| (J) Leveling of seafloor or embankment | Hydraulic "dustpan" dredge with heave compensator suspension of dredge head, drags, bottom-supported screed frame, screed frame from TLP or heave-compensated platform, horizontal screw auger. |
| (K) Provision of uniform support under base of structure | Preleveling as per J (above), underbase grouting, underbase sand injection or sand flow, tremie concrete, grout-intruded aggregate ("Prepakt"), mud jacking. |
| (L) Excavation beneath structure | Articulated dredge arms, airlift, jets, eductors, drills. |
| (M) Scour and erosion protection | Sacrificial fill, riprap, graded stone with filter course, filter fabric, articulated mattresses, sandbags, grout-filled porous bags, skirts on structures, aprons and flow-control devices at base of structures, artificial seaweed, sand-asphalt and stone-asphalt blankets, underwater concrete slabs. |
| (N) Turbidity suppression | Betonite-cement slurries, discharge of fine sand blanket, bucket deposition of clay, flocculants. |

If the cement can be mixed with the soils, even cohesive soils can be stabilized. Various techniques have been developed to accomplish this, usually based on use of an auger-type drill that is jetted and drilled into the clay. The jet water is then turned off and a thin cement slurry injected, to be distributed by the mechanical action of the auger drill. The Japanese have developed several such methods and have applied them to shallow seafloor soils. Jet grouting and deep cement mixing appear to be the most efficient systems for stabilizing weak clays and silty sands.

Another method is based on the injection of cement slurry under relatively high pressure into clays and silts. The soil is first fractured, by high-pressure water, allowing multiple lenses of cemented material to become interbedded, thus increasing the soils shear resistance. The method can be extended to remove the weak soil from a drilled shaft and replace it with grout.

A similar method of injection uses quicklime (CaO) which is placed in an augered hole in a watertight container to avoid premature hydration; the drill is then withdrawn, with sand being used as a packer on top of the lime; then the drill ruptures the watertight container. The quicklime draws water from the surrounding soil and is converted to calcium hydroxide, with significant liberation of heat. Stable calcium compounds such as calcium carbonate are formed among the clay particles.

Use of chemicals other than limes and cements is also practicable in permeable and slightly permeable soils. For many years, injection of sodium silicate followed by injections of calcium carbonate has been used for stabilization. In some calcareous soils, if they are sufficiently permeable, a single-stage injection of sodium silicate may be adequate, being fixed by the free lime in the soil.

Organic polymers may also be injected, either as polymers or as monomers which will later convert. These monomers are usually highly toxic, and hence their effect on adjacent marine life must be considered. Some of these have high penetration qualities which make them very effective in low permeability clays and silts. Shell Oil International has developed a penetrating polymer known as Eposand. The soil is first flushed with fresh water, then with an organic solvent, following which the Eposand is injected.

*Chapter 5*

# *Moorings and Anchors Towing and Diving*

## Introduction

Vessels working at an offshore site must be held in position despite the effects of wind, waves, and current. The current forces are relatively constant in direction in the offshore zones, in closer-in areas and opposite the mouths of great estuaries they may vary with the tidal cycle. The wave forces can be considered as comprising an oscillatory motion plus a steady, slow drift force. Both the mean forces of a quasi-static nature and dynamic forces must be resisted.

The standard means of mooring is by way of a mooring system that connects the vessel (or structure) to the seafloor by means of laterally leading lines to anchors. Moorings must be thought of as a system which includes the vessel, the anchor engines, fairleads, mooring lines, buoys, and anchor. In deep water they can be of the catenary type, extending from the vessel in a catenary to the seafloor and thence laterally to the anchor. In shallow water, taut moors may be employed in which the mooring lines are tensioned to run relatively straight from the vessel to the anchor or fixed structure. Recently, taut moors have also been used in deep water.

Manned intervention in underwater construction is one of the oldest forms of offshore activities, dating back at least as far as the ancient Romans, Phoenicians, and Indians and probably to even older civilizations. As practiced in the offshore today, diving and the use of highly skilled technicians in the underwater environment are a very advanced technology, supported by extensive research and development in such disciplines as physiology, psychology, communications and control, power systems, and mechanical devices. Underwater tools and electronic-acoustic systems have greatly enhanced the effectiveness of divers.

Now let us have a look at Towing.

Certain basic principles apply to towing. One is that the attachments to the structure or barge must always be sufficiently strong that they do not fail or damage the structure under the force that parts (breaks) the towline. The actual breaking strength of wire rope is typically 10 to 15 per cent greater than the guaranteed minimum breaking strength. Actual breakage will usually occur under a dynamic load rather than a static load. It is important that under overload, the structure or vessel being towed remain undamaged. A usual requirement is that the ultimate capacity of any towline attachment to the unit be at least four times the static bollard pull and at least 1.25 times the breaking strength of the towline from the largest tug to be used on that attachment. At least one spare attachment point, with pennant, should be fitted for towing ahead, to be used in case of emergencies.

A second principle is that the towing force must be able to be resisted through a significant range of horizontal and vertical angles, thus imparting shear and bending, as well as tension, on the towing attachment.

If a towline does break at sea, it is desirable that it fail at a known "weak link" so that it may readily be reconnected, even in high sea states. A typical arrangement when a single boat is towing with a bridle. If the towline is subjected to a high-impact overload, the short pendant between B and C breaks, the shackle at B is pulled back on deck by means of a fiber rope pendant, a new pendant fitted (BC), and the towline reconnected. To reduce shock loads in the towline, either a highly elastic fiber pendant or a length of chain may be used.

During passage through restricted waters and during final positioning, the towline may be shortened in scope to permit better control. If it is too short, however, the thrust of the propeller's wash will react against the towed vessel. When one of the large GBS caisson structures was being moved in Stavanger Fjord, Norway, the lines had very short scope in order to control movement between rock islands. The thrust of the propeller wash against the 120-m-wide and 50-m-deep projected area of the caisson resulted in inability to get the structure to move. The solution was to place the primary tugs at the rear of the caisson, pushing in notches fabricated of steel and timber. Thus the full efficiency of the propeller's thrust could be developed.

## Moorings and Anchor System

The mooring system that connects the vessel to the seafloor by means of laterally leading lines to anchors, is considered as the standard means of mooring.

Some large semisubmersible drilling vessels and drill ships use chain mooring lines, with winches specially fitted for this use. Longer-term moorings may use a hybrid combination of steel wire and anchor chain.

The sheave diameter of fairleads should be at least 20 times the diameter of the wire rope. When mooring lines break, they usually do so at the fairlead, for this is where bending stresses are added to direct tension.

Anchors are of a number of basic types. First are the reusable drag anchors, which have evolved from ship anchors; they include the stockless type used by the U.S. Navy, the Danforth, and the newer Bruce, Stevin, and Doris anchors. These anchors are designed so that, as a horizontal force is applied, they dig down into the soil and mobilize it as resisting force. They are often rated on the multiple of their holding power to their air weight. This is an oversimplification, since it is the soil which they must mobilize, and the resistance varies with the characteristics of the soil and the configuration of the anchor. Some anchors are specially designed for soft clays, others for sands.

All these anchors require that the pull be horizontal. In fact, they are purposefully designed so that a vertical pull breaks them loose with little more force than their weight. This means that the portion of the mooring line immediately ahead of the anchor must be heavy enough to stay seated on the seafloor even when the line is under full tension. One or one and one half shots of chain are usually

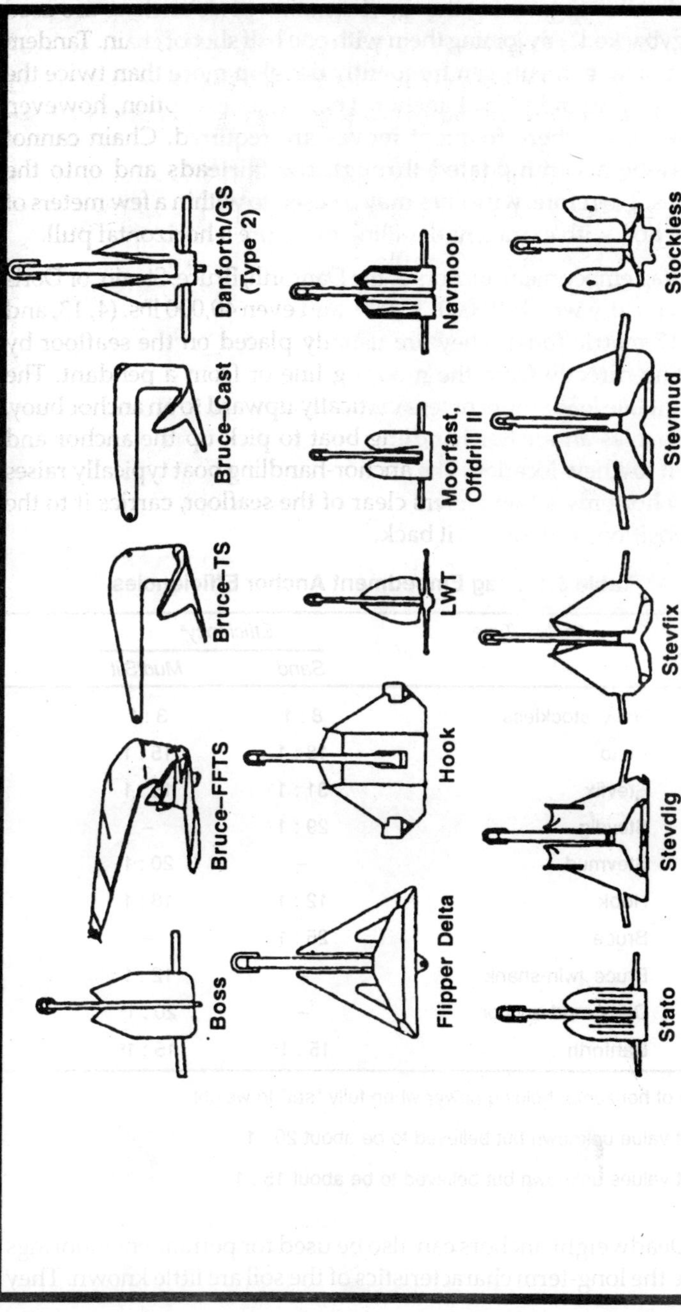


**Figure 5.1: Drag Embedment Anchors**

placed in this segment of the line. Sometimes two anchors are used ("piggybacked") by joining them with one half shot of chain. Tandem anchor arrangements can frequently develop more than twice the capacity of an individual anchor. There is an exception, however, for the cases where frequent moves are required. Chain cannot usually be accommodated through the fairleads and onto the winches. Therefore, wire lines may be used to within a few meters of the anchor, with extra length of line to ensure a horizontal pull.

Drag embedment anchors of the Danforth, Bruce, Stevin, or Doris type typically weigh 10,000, 30,000, and even 40,000 lbs. (4, 13, and even 17 metric tons). They are usually placed on the seafloor by lowering directly from the mooring line or from a pendant. The pendant line leads more or less vertically upward to an anchor buoy. This enables an anchor-handling boat to pick up the anchor and move it to a new location. The anchor-handling boat typically raises the anchor only a few meters clear of the seafloor, carries it to the new position, and lowers it back.

**Table 5.1: Drag Embedment Anchor Efficiencies**

| *Anchor Type* | *Efficiency*[a] | |
|---|---|---|
| | *Sand* | *Mud/Slit* |
| Navy stockless | 8 : 1 | 3 : 1 |
| Stato | 18 : 1 | 15 : 1 |
| Stevfix | 31 : 1 | 15 : 1 |
| Stevdig | 29 : 1 | – |
| Stevmud | – | 20 : 1 |
| Hook | 12 : 1 | 18 : 1 |
| Bruce | 25 : 1 | – |
| Bruce twin-shank | – | 12 : 1 |
| Doris mud anchor | – | 20 : 1[b] |
| Danforth | 15 : 1[c] | 15 : 1[c] |

[a] Ratio of horizontal holding power when fully "set" to weight.

[b] Exact value unknown but believed to be about 20 : 1.

[c] Exact values unknown but believed to be about 15 : 1.

Deadweight anchors can also be used for permanent moorings where the long-term characteristics of the soil are little known. They

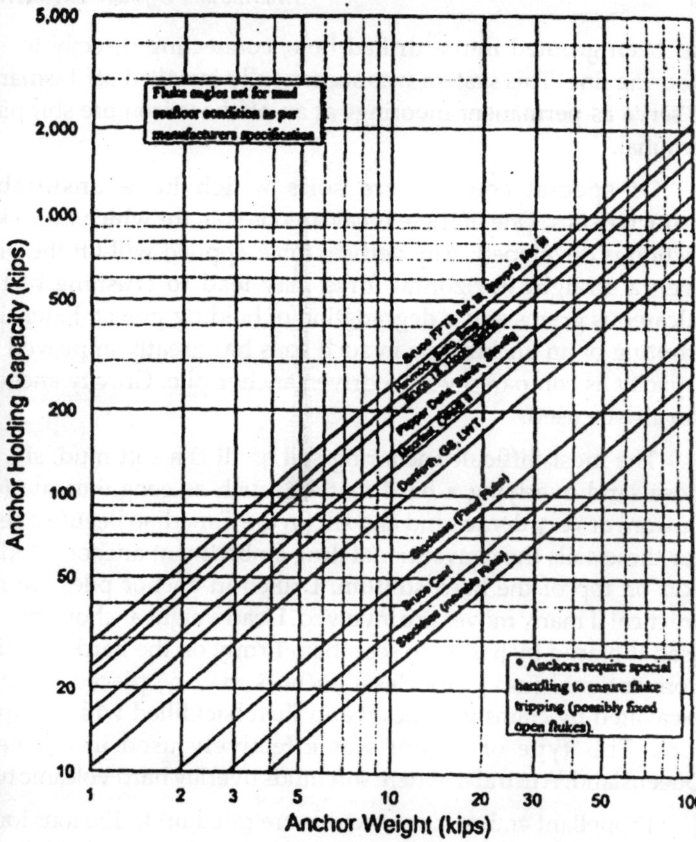


**Figure 5.2: Anchor System Holding Capacity in Soft Clay**

can be used where the direction of pull changes radically from time to time since they are omnidirectional, whereas the Danforth-type anchor will just pull out when the pull reverses direction.

Pile anchors are very effective in many soils. The pile can either be drilled in and grouted, using an offshore mobile drilling rig, or driven in with an underwater hammer or a follower. The anchor line, usually a shot of chain at this location, can lead from the top or from a point a few meters down the pile. The anchor pile resists pullout by a combination of bending plus passive resistance (the P/y effect) and skin friction shear. In some cases, in rock, a chain has been grouted into a drilled hole, connecting directly to the mooring line. This system was successfully installed off Tasmania to serve as permanent moorings at an offshore iron ore shipping terminal.

Of special concern are soils which have unsuitable characteristics. One of these is calcareous soil, for which little skin friction is developed. Any vertical force applied will lift the pile. Even a straight horizontal force may lead to crushing of the calcareous grains and a degradation of holding power. Extensive grouting of an anchor pile in such soils has greatly improved its capacity as compared with a driven anchor pile. Gravity anchors can also be used.

The most difficult anchoring soil of all is a soft mud, silt, or loose sand overlying a hard material such as conglomerate (off Taiwan) or very dense sand and silt (in the Canadian Beaufort Sea). For these soils the conventional drag embedment anchors tend to skid on top of the hard stratum. Drilled-in anchor piles are not practical if many moves are involved. Deadweight anchors may be used if placed by jetting to seat them firmly on the hard material. Conventional (navy stockless) anchors may be placed in holes excavated by clamshell bucket and then backfilled with dumped rock. This type of anchor was effectively used in northern Queensland, Australia, where soft muds overlay hard volcanic tuff.

Propellant embedment anchors are rated up to 150 tons long-term capacity in soft mud and clay soils. Actual values are higher in sand and coral, ranging up to 300 tons. These anchors are multidirectional, are installed rapidly, and function best where drag anchors are least effective.

For use in the deep sea (over 200 m), the anchors must resist primarily vertical forces. Very heavy concrete deadweight anchors may be used. Recent developments include drag anchors shaped to develop high vertical capacity. They are seated by horizontal pull, then rotated (or the flukes rotate) to resist uplift.

Suction anchors gain their vertical capacity by skin friction on the inner and outer surfaces. These are large-diameter (greater than 5 m) steel cylinders, perhaps 20 to 30 m in length which are capped by a steel dome. Hydraulically operated valves in the dome permit opening or closure.

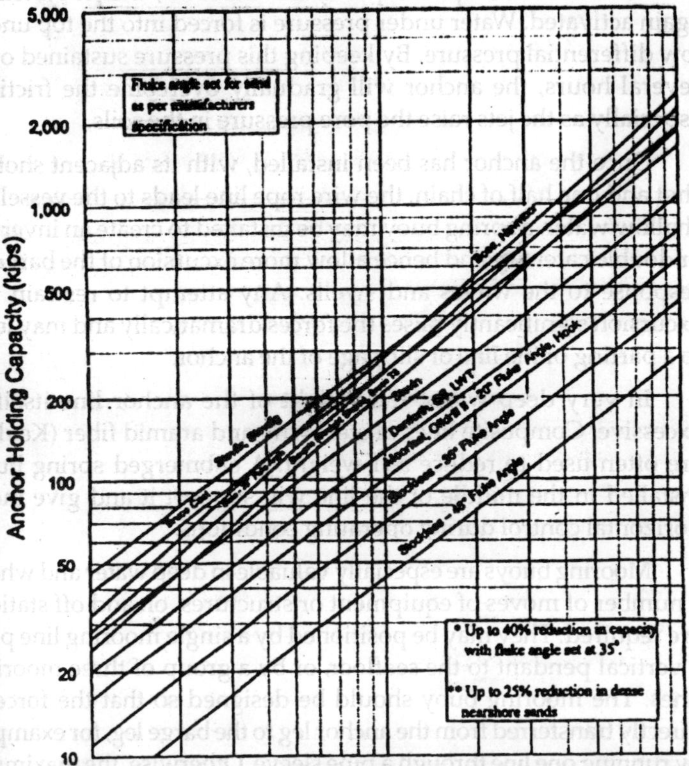


**Figure 5.3: Anchor System Holding Capacity in Sand**

The suction anchor is lowered to the seafloor with top valves open, and allowed to penetrate under its own weight. In some soils, jets may be operated to increase penetration, using low pressure and flow rates so as not to produce piping. Then the top valves are closed and the water pumped out to reduce the pressure under the dome. This gives a driving force equal to the differential hydrostatic pressure, and forces the anchor deeper into the soil. The resistance to uplift is now the skin friction shear on the inside and outside periphery, with a safety factor furnished by the fact that any slight upward movement produces a temporary added resistance due to the suction. Suction anchors typically develop several hundred tons of both lateral and uplift capacity. For removal, the jet system is again activated. Water under pressure is forced into the top under low differential pressure. By keeping this pressure sustained over several hours, the anchor will gradually overcome the friction, especially as the jets raise the pore pressure in the soils.

Once the anchor has been installed, with its adjacent shot or shot and one half of chain, the wire rope line leads to the vessel. In shallow water, a spring buoy may be installed to create an inverted or double catenary and hence allow more excursion of the barge in response to the waves and swells. Any attempt to restrain the excursion significantly raises the forces dramatically and may lead to a parting of the line or slippage of the anchor.

In very deep water, the weight of the anchor line itself is excessive. Composite lines of steel wire and aramid fiber (Kevlar) are often used to reduce self-weight. A submerged spring buoy installed in the middle of the line will support it and give more horizontal control during operating conditions.

Mooring buoys are especially valuable in deep water and where a number of moves of equipment or structures, on and off station, are required. They may be positioned by a single mooring line plus a vertical pendant to the seafloor, or by a group of three mooring lines. The mooring buoy should be designed so that the force is directly transferred from the anchor leg to the barge leg, for example, by running one line through a pipe sleeve. Otherwise, the maximum forces may damage the buoy. The buoy should be designed to resist the maximum hydrostatic head if it is pulled underwater. It should be foam-filled. Many unfilled buoys have been sunk by rifle shots from fishermen. The Mini-OTEC riser buoy was not filled: it was sunk by a rifle shot and the entire riser as lost.

Although a mooring buoy may be held in approximate position by a vertical pendant to a clump anchor, to hold it in a fixed position requires three mooring lines. For offshore work, the effective pull which the buoy must resist is usually directional, within a spread of 30 to 45°. Therefore, two legs of mooring lines are led out to anchors. The third leg is a short leg, either attached to a clump anchor more or less directly below the mooring buoy (it will never be directly below) or leading with short scope toward the mooring position.

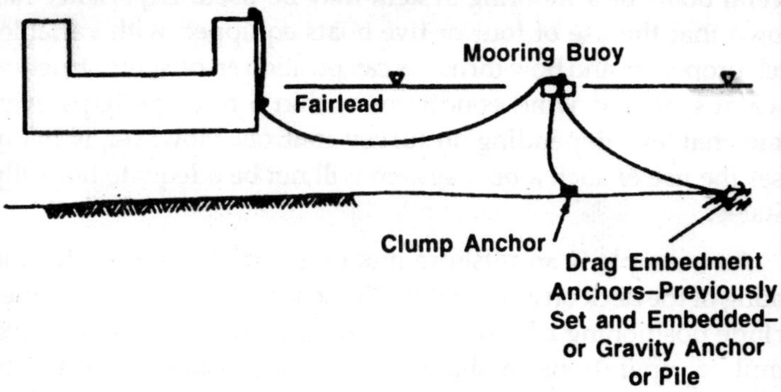


**Figure 5.4: Mooring Line Arrangement**

For very large structures such as gravity-base platforms having long response periods and where the moorings have resilience to absorb shock loads, the design should be based on the significant wave plus a 1-min sustained wind at the relevant heights and maximum current. Allowance may be made at sheltered inshore locations for reduced wind and waves and the changed geometry of the mooring due to excursion under load.

The maximum load in a new or used mooring chain with its associated shackles and fittings, including residual pretension, should not exceed 70 per cent of the minimum breaking load, after allowance for corrosion and wear. Note that many manufacturers and classification societies quote an average breaking load rather than the minimum. Where wire line moorings are used, the maximum load should not exceed 60 per cent of the guaranteed minimum breaking load.

The design holding capacity of any anchor, winch, or connection, multiplied by appropriate safety factors to account for gusts and dynamic amplification, should exceed the extreme storm loading on it. Note that the effect of long-period response motions in sway, surge, and especially yaw may impose significant increases in force. The attachment points to the structure should be designed to resist at least 1.25 times the nominal breaking load of the mooring, without damage to the structure.

For positioning a large offshore structure in the open sea, either several boats or a mooring system may be used. Experience has shown that the use of four or five boats equipped with variable-pitch propellers and bow thrusters can position an offshore structure in calm seas and wind conditions within 5 m or perhaps even somewhat less, depending on survey controls. However, in many cases the use of such a boat system will not be adequate nor fully suitable.

An example is an offshore mooring system required for the mating of the deck on a concrete GBS substructure. Other examples include positioning a bottom-founded structure over a predrilled template, operations in shallow water, and those subjected to significant currents. An offshore mooring system is also required when the structure or vessel must remain on location for a significant period of time, subject to changes in the sea state and wind conditions.

In shallow waters, off the coast of Queensland, Australia, 10 gravity-base caissons were required to be positioned within 0.5-m tolerance. This was achieved by use of preset mooring buoys. The arrangement adopted, using a spring mooring buoy, gave upward spring action, with the buoy serving as a spring buoy during final setting, thus absorbing shock and dynamic surge loads. This system is also called an "inverted catenary."

For mooring of an offshore derrick barge working in coastal waters, two systems have been found necessary. On the San Francisco

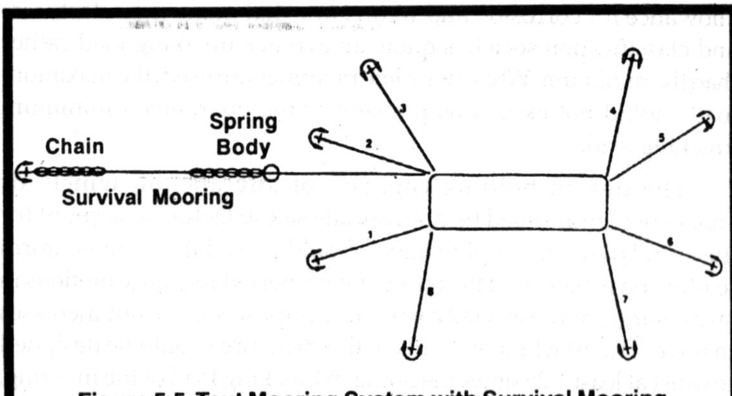


Figure 5.5: Taut Mooring System with Survival Mooring

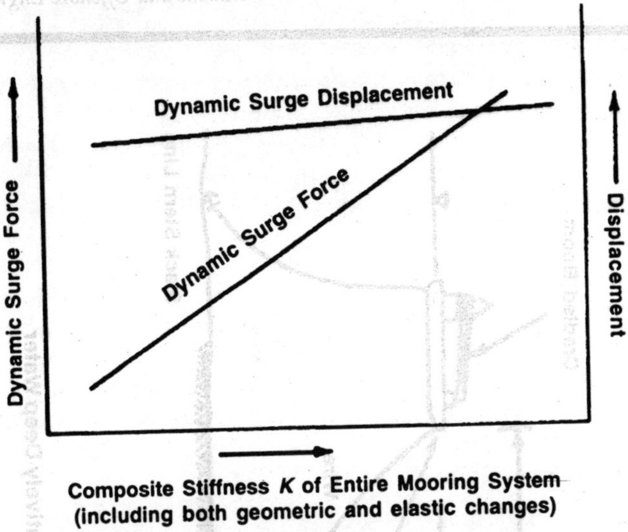


**Figure 5.6: Effect of Stiffness of Mooring System on Maximum Forces and Displacements**

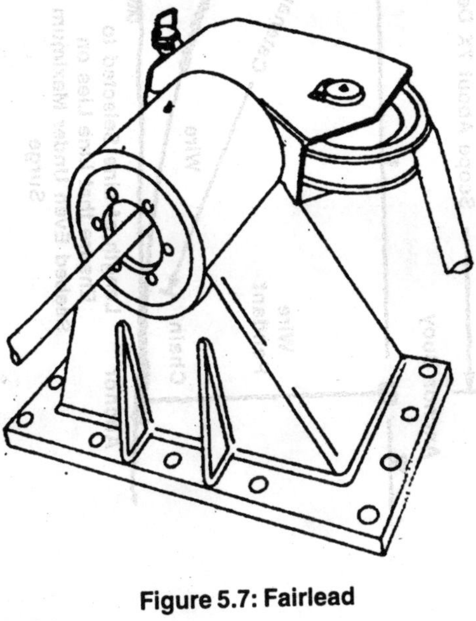

**Figure 5.7: Fairlead**

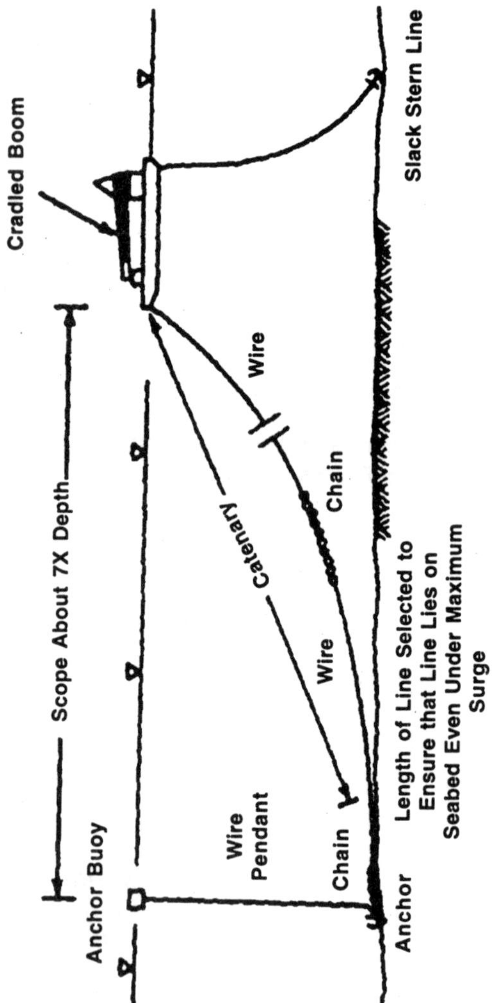


**Figure 5.8: Survival or Storm Mooring in Relatively Deep Water**

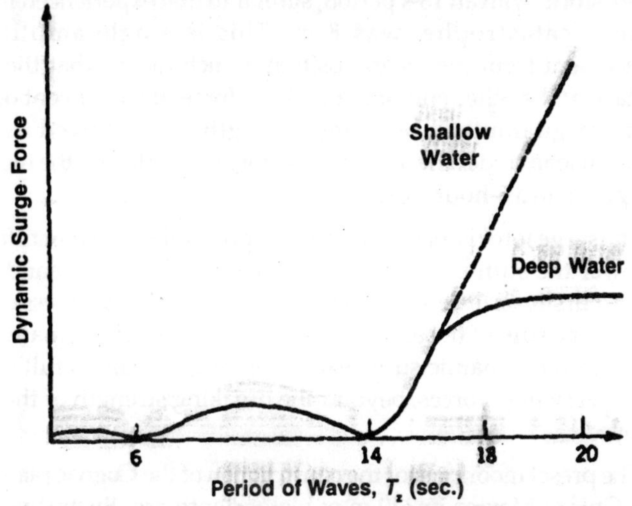


**Figure 5.9: Dynamic Surge as a Function of Wave Period, for Both Deep and Shallow Water**

Southwest Ocean Outfall Project, a 4-m-diameter buried concrete pipe is being installed 7000 m into the sea in water 10 to 30 m deep. An offshore derrick barge was initially moored on a taut mooring in order to enable it to carry out operations requiring accuracy and control. An intense storm with unpredicted long-period waves created enormous surge forces that broke the taut mooring lines and drove the vessel onto the beach, severely damaging it. As noted earlier, long-period waves in shallow water develop elliptical particle orbits and create shorter and steeper waves resulting in increased surge accelerations. When the offshore barge was returned to the site, the taut mooring system was again used for operations, but in addition a survival mooring system was installed.

If the water had been deeper, the survival mooring would have been configured as a catenary, perhaps with a clump or chain in the bight to provide greater spring action. However, in shallow water the stretching out of the catenary permitted very little movement in surge. Hence, other means had to be taken. In this case, both a spring buoy, giving geometric travel, and a very long wire line, giving elastic stretch, were employed. The dynamic surge to be encountered during a severe storm with an 18-s period, similar to that experienced in the previous catastrophe, was 8 m. This is single-amplitude displacement from the mean position, which means that the line will slack at the other end of the cycle. A force of 65 per cent of the minimum guaranteed breaking strength was allowed as the maximum load under the design surge force imparted by the highest wave group in a 6-hour storm.

The surge force is, of course, directly proportional to the stiffness. The lower the stiffness, the less the force. On the other hand, the surge excursion is little affected by the stiffness. The stiffness of the system is the sum of the elastic and geometric stretch. Any attempt to restrain the dynamic surge excursion to less than its full value leads to very high forces, beyond the breaking strength of the line employed.

The preset moorings for the construction of the Cognac platform in the Gulf of Mexico in 320 m of water depth are illustrative of a major installation carefully engineered to ensure that the operations could proceed without difficulty. A 12-point mooring system was adopted. Each of the 12 mooring buoys was built of steel, filled with 2 lb/ft$^3$ (30 kg/m$^3$) closed-cell urethane foam, and equipped with controllable light signals for guidance of the workboats so they would not run up on a taut line. The buoys were then anchored by 400 m of 2¾ in. (68 mm) chain, which led to pile anchors. The pile anchors were 30 m long, 0.75 m diameter, with 25 mm walls. They were jetted into place by a drilling vessel. From the buoys, two parts of 50 m (2 in.) wire rope line led to the barge. Between the two derrick barges, four 112 mm-diameter (14-in.-circumference) nylon lines were run in order to maintain relative position yet absorb dynamic loads arising from differential sway and yaw of the barges.

Dynamic positioning "thrusters" are being increasingly employed to maintain position of the construction vessels on some or all of the axes and thus fulfill part or all of the mooring requirements. These thrusters may be mounted on deck or within the hull. In their most-sophisticated arrangements, they are controlled by minicomputers and GPS and utilize variable-pitch propellers so that they run at constant speed.

## The Installation of Marine Structures

The installation of marine structures usually includes the lifting and setting of modules and other heavy loads on the platform. Such lifts may weigh up to 2000 to 4000 tons and more. Loads up to 13,000 tons have been set by derrick barges with two cranes. The hammerhead crane SVANEN has been successfully employed to erect piers, shafts, and girders up to 8000 tons for very long bridges in Denmark, Sweden, and Eastern Canada. For the 24,000-ton prefabricated piers of the Oosterschelde Storm Surge Barrier, 12,000 tons was supplied by buoyancy and 12,000 tons by a catamaran lift barge. Installation involves motion and hence dynamic loading and impact effects. API RP2A, sec. 2.4. "Installation Forces," recommends specific precautions to ensure safety in the handling of such loads. The DNV Rules, appendix H–1, "Lifting," specifies procedures and rules to ensure safe lifting of heavy loads at sea.

When lifting a heavy load, there are both static and dynamic forces to consider. The static force include the actual load itself, which if not weighed, must be computed to include the design weight, plus adequate allowances for over-tolerance plate thickness, weld material, padeyes, and any supplies stored within. Static lifting loads must also include the slings, spreader beams, and shackles.

The author once investigated a critical lift that was at the limit of rated capacity of the 500-ton crane barge. A careful physical inspection revealed that over 50 tons additional of tools and supplies had been stored aboard by the drilling crews. Worse, many of these, including an acetylene bottle, were loose, that is, not properly secured.

The dynamic forces are those due to acceleration, first as the load line lifts while the load, still resting on the barge, is starting the down-heave cycle. Later, both horizontal and vertical accelerations are imposed during swing. Lifting forces on the padeyes and the structural members of the load to which they are secured have both vertical and horizontal components. Many modules are designed to withstand the vertical quasi-static forces imposed in lifting in the fabrication yard, where bridge cranes or skids may be employed. At sea, however, the lead of the slings is usually inclined in two planes. Although the padeyes themselves are usually adequately designed for vertical and horizontal loads, the structure to which the padeye connects must also be able to accept and transmit the total vertical and horizontal forces back into the structure.

Modules fabricated in Houston or Singapore are initially lifted in warm weather. When later lifted at the site, which may be the Bering Sea, cold weather impact properties become important.

Vertical forces on lifting can include the favorable effects of buoyancy where applicable; however, fully or partially submerged structures may pick up an added hydrodynamic mass component. This latter may be a very high factor when the submerged surface is horizontal. For example, a proposal, was made by one offshore contractor to lower a 50,000-ton box-like unit to the seafloor of the North Sea. He proposed to utilize the buoyancy of the box to reduce the net weight to a few hundred tons, well within the rated capacity of the derrick barge. What he failed to recognize was that due to added mass effects and inertia, the rectangular box-like unit would react almost not at all to a 6- to 7-s wave, whereas the derrick barge would have significant pitch and heave during that same period. While the stretch of the lines would have accommodated some of the dynamic effects, a detailed analysis showed that the derrick booms would have been seriously overloaded.

Instrumentation is now available to enable control of the dynamic aspects of lifting. These consist of sensors on the crane barge, on the crane boom, and on the barge or boat from which the module or other lift is being lifted. Typically, mini- or microcomputers then give readouts of "load on hook," "out reach" (radius), "hook height," "wave height," "wave period," "derating of crane capacity for sea state," "hook speed," "net load on deck of crane barge and effect on stability," "crane hook height," "off-lead" (distance between load and fixed structure), "automatic level luffing," and "warning as to turns remaining on winch drum." Other programs are available to determine optimum heading of crane barge to minimize boom tip motion and hence the dynamic increment of load during the operation.

An example of a very heavy lift was the 5400-ton deck of the *Esmond* topsides. It was carefully engineered for setting with Heerema's derrick barge *Balder,* with its two cranes rated at 2700 and 3000 tons. Positioning was achieved using conventional taut mooring lines, along with a computerized ballasting system for the crane barge. The entire operation took only 1 hour.

Another example was the 10,700-ton integrated deck for the Britainia platform in the North Sea. This was set by the twin-crane barge *Thialf*, which has a nominal capacity of 12,000 tons. The barge is 200 × 88 m, and is moored by 12- to 22.5-ton anchors, using 80 mm high-strength steel lines, plus six dynamic azimuth thrusters.

A very heavy lift crane vessel, *Rambo 12*, was built up by joining two heavy-duty crane barges into a catamaran. The two barges were linked by a third barge arranged transversely between the bows of the two barges and thus enhanced both transversal and longitudinal stability. The two cranes, each of 2000 tons capacity, were linked to make 3000-ton lifts for the Tagus River Bridge in Portugal.

For the Second Severn Bridge in England, two large skid-mounted cranes were installed on a large jack-up barge. They lifted the 1500-ton pier shells for the foundations.

The Svanen is a giant hammerhead crane which was specially built to place the pier shells, pier shafts, and girders of the Great Belt Western Bridge. It lifted 330 major precast segments weighing 7000 tons each. Then it was modified and transported by submersible ship to Eastern Canada where it lifted and placed piers with weights up to 8000 tons each, as well as pier shafts and girders of the Confederation Bridge linking New Brunswick and Prince Edward Island. Then it came back across the Atlantic to participate in the Øresund Bridge joining Denmark and Sweden.

Catamaran lifting barges have been used for the transport and control of many submerged tubes, with net weights of 1200 tons. Recently they have been proposed for lifts of 3000 tons on lock and dam construction projects.

Lifting eyes are designed to transmit the load to the slings in the plane of the sling. As the structure swings, however, or the barge from which it is being picked sways, a side loading may be imposed. API RP2A recommends that a horizontal force equal to 5 per cent of the static swing load be applied simultaneously with the static swing load. It is to be applied perpendicular to the padeye at the center of the pin hole.

When suspended, the lift will assume a position such that the center of gravity of the load and the centroid of all upward-acting forces are in static equilibrium. These relative positions should be taken into account in determining the inclination of the slings. The force in the sling is the resultant of the horizontal and vertical forces at the padeye, as computed for the most severe inclination of the

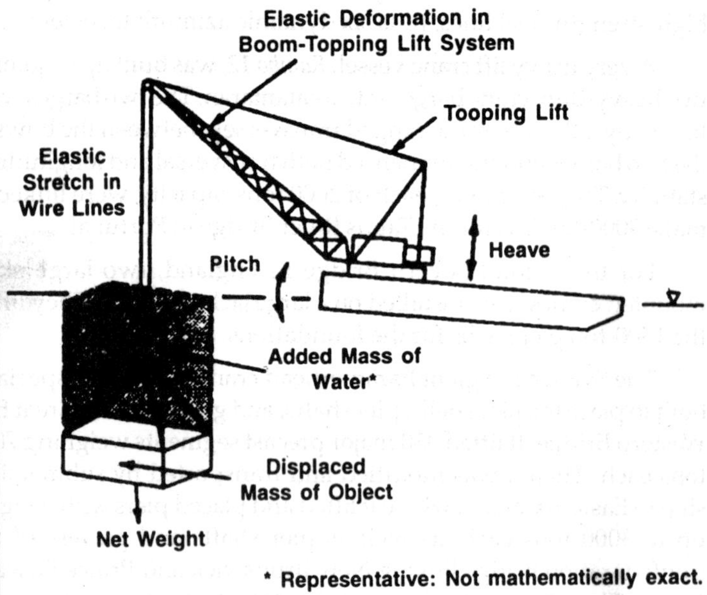


**Figure 5.10: Schematic Representation of Dynamics of Lowering Loads Under Surface of Sea**

sling. Due to swinging of the load while in the air, the load will not be uniformly distributed on all four slings. This nonuniform distribution must be considered in sizing of the slings and their fittings.

As the load is picked, and again as it is set, the position may vary from the above, due to the horizontal and vertical reactions from the deck of barge or platform, as well as those from tag lines and guides. The change in horizontal and vertical forces so occasioned must be considered in determining the forces and angles of application on padeyes and hooks.

API RP2A recommends that for lifts to be made in the open sea, a minimum load factor of 1.35 should be applied to the calculated static loads. This must then be multiplied by the material factor of 2.0. Thus it is a somewhat less than conventional wire-rope rigging design, for which a factor of 4 to 5 is normally applied to the minimum guaranteed breaking strength to determine the safe calculated static load. However, it is suitable for padeyes and structural members.

The above factors should also be applied to the padeyes and other internal members connecting directly to the padeyes. All other structural members transmitting lifting forces within the structure should be designed using a minimum load factor of 1.35. For lifting in cold weather (below +5°C), adequate Charpy impact valves should be verified.

OSHA requires a factor of safety of 5 on ultimate strength on all lifting gear (wire rope shackles and padeyes), whereas ANSI requires a factor of 3 on yield strength. The two codes are roughly equivalent. During load-out in sheltered harbors, where there are no waves, the load factors can be reduced, but not less than 1.5 and 1.15, respectively. Note that such reduced factors may be technically in violation of local rules and thus require waivers.

Structural members, padeyes, and other attachments for lifting of lighter members in harbor and bridge construction are usually designed on the basis of allowable (elastic) stresses, using an approximate factor for impact (usually 2.0), with no increase in stress allowed because of short-term loading. The allowance for impact by a factor of 2.0 is also applicable to lifting inserts in concrete. In addition, all critical structural connections and primary members should be designed to have adequate ductility to ensure structural integrity during lifting even if temporary or local overloads occur. Special attention must be given to ensuring weld ductility and prevention of undercutting and adverse heat affects on the surrounding metal (HAZ). Low-hydrogen electrodes should be used. The design of padeyes requires special attention and detailing. Given the forces, including dynamic and range of angles in both planes over which the forces may act, the padeye must transfer the load from the pin of the shackle into the structural frame.

Transverse welds perpendicular to the principal tension, where the member is subjected to impact, are prohibited by some national codes. If they are used, the details, welding procedures, and non-destructive testing (NDT) used to verify them must be such as to ensure full development of ultimate strength and ductility.

Wherever possible, the load from the padeye to the girder should be transmitted in shear. The use of cheek plates should be avoided. The distance from girder to pin should be at least six plate thicknesses.

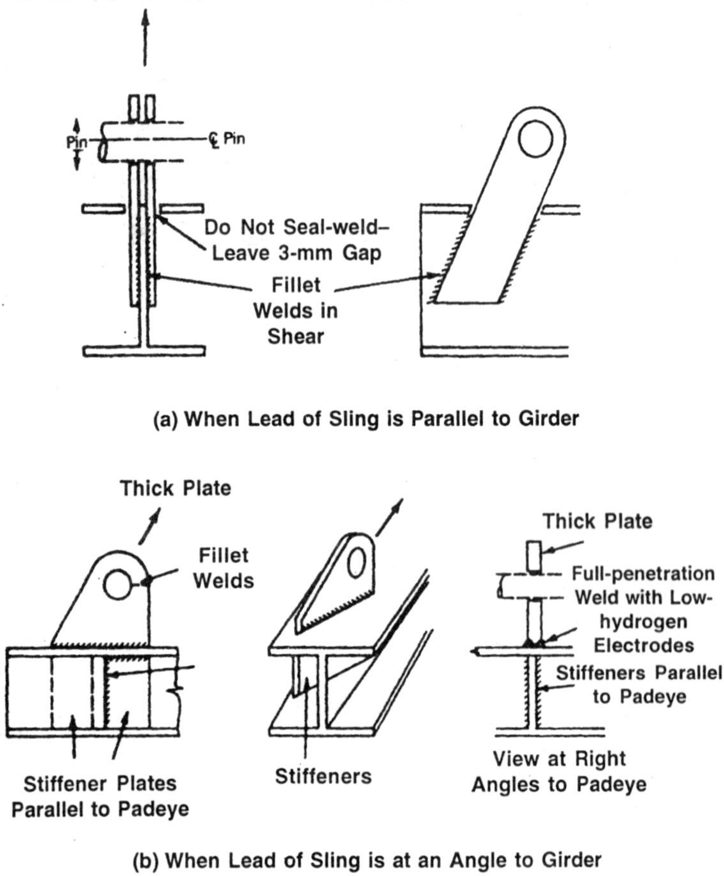


**Figure 5.11: Satisfactory (Safe) Padeye Details**

Where cheek plates are necessary, they should be welded to the main load-carrying plates with bevel welds, sufficient for an even transfer of stresses, on the inside and with fillet welds on the outside. The bore of the cheek plates should be flush with the bore of the padeye, to achieve uniform bearing on the shackle pin. This may require reaming after welding. The direction of rolling of the parent steel plate should be determined, and this direction should then correspond with the direction of the sling.

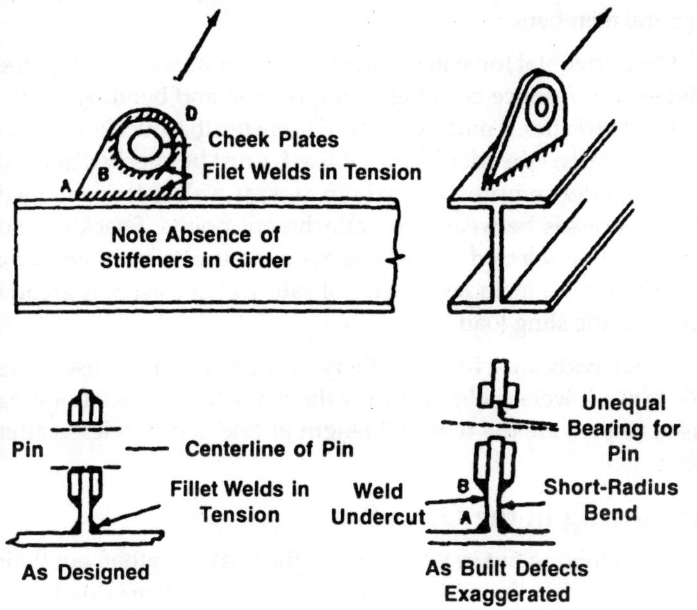


**Figure 5.12: Example of Incorrect and Dangerous Padeye Details**

If connection between padeyes and structure cannot be transmitted by shear, then full-penetration welds can be employed, using the welding procedures prescribed for the primary structural members, with full NDT inspection. The structural plate to which the padeye is attached must have adequate through-thickness toughness to prevent laminar tearing.

API RP2A calls attention to the fact that fabrication tolerances and variations in sling lengths can redistribute the actual forces and cause significantly increased stresses in some members. The variation in sling length should not exceed to ±0.25 per cent of nominal sling length or 37 mm. The total variation from the longest to the shortest sling should not be greater than 0.5 per cent of the sling length or 75 mm.

Where unusual deflections or particularly stiff structural systems are involved in the lifted load, a detailed analysis should be made to determine the redistribution of forces of both slings and structural members.

The horizontal force in the structural members connecting the padeyes can produce both high compression and bending due to minor eccentricities. It must be checked to ensure that buckling cannot occur. Similarly, spreader beams, if used, must be checked against buckling under compression on both axes, as well as for combined bending stresses between sling attachment points. Shackles and pins should be selected so that the manufacturer's rated working load, provided it includes a factor of safety of at least 3, is greater than the static sling load.

Lifted loads must be controlled against swinging by use of tag lines. When lowering objects below the sea surface, special rigging is used to provide the required length of line and to release after setting.

## The Towing in the Open Sea

When towing out in the open sea, the boats lengthen out their towlines to offset the wide range of loads in the lines due to the waves and swells. When towing a very large structure in coastal waters, a single lead boat may run ahead to verify route, confirm depths by forward-looking sonar, and pick its way through underwater obstructions, or ice. Such a lead boat can also warn other shipping, which may be disregarding published Notices to Mariners.

If the towed structure is a deep-draft vessel (some of the offshore platforms in the North Sea have drawn 110 to 120 m), then the towline, if attached to the structure below water near to the center of rotation, may have a steep inclination. This will tend to pull the stern of the boat down into the water. Therefore, the towline may be led up to a pontoon or buoy, which will resist the vertical component of the towing force. Such a buoy should be foam-filled to prevent flooding in event of a leak or hole. Such a system may also be useful when towing through broken ice to minimize the shock loads in the towline itself.

In the case of the Dunlin platform and the Ninian Central platform, flotation units shaped like sausages were clamped to the towlines. These floaters were filled with polyurethane, each giving approximately 5 tons net buoyancy.

Most channels, harbors, and shallow offshore coastal areas have been extensively surveyed, with the results published on hydrographic charts. Unfortunately, the depths of interest were those which pertained to ships having drafts of 10 to 20 m. Hence, when the survey ship got into deeper waters, it usually only recorded depths on a grid, with no interest in determining possible rises, shoals, or pinnacles as long as they did not present a hazard to shipping.

A similar problem arises when towing vessels or structures in areas not normally used by shipping. Hence, careful surveys need to be made, using sonar, side-scan, and profiler acoustic equipment so that both the route and its full swath, including sway excursions, are thoroughly scanned.

Required channel widths in sheltered areas are usually twice the beam, but this must be considered in relation to the environmental conditions and navigational accuracy. For exposed areas, the required width will depend on currents and navigational accuracy and thus may vary from about 600 to 1500 m for relatively short distances of 12 km or so.

During the tow between islands or shoals, accurate electronic position-finding systems need to be set up. Unlike the case of a towed ship or barge having a draft of perhaps 8 to 10 m and a width of 30 to 40 m, an offshore structure such as a deep-water caisson may have a draft of over 100 m and a width of 100–150 m.

Therefore, it is not enough to plot only the position of the "bridge"; the extremities must also be charted. Detailed current surveys, both surface and at depth, must be made in restricted areas.

A structure under tow will experience sway and wander somewhat on its course. In confined waters, a band may be plotted, shaded in color, within which the structure is safe. Then as the edge of the structure approaches the band edge, corrective action can be taken. This will eliminate excessive "hunt" back and forth trying to stay exactly on a course line.

Towed vessels and shallow-draft structures may have an actual draft greater than their mean draft. This may be due to trim, squat, list, or wind heel. It may be due to the lower density of fresh water discharging from a river into the adjacent sea: fresh water reaches long distances from the mouths of such rivers as the Orinoco,

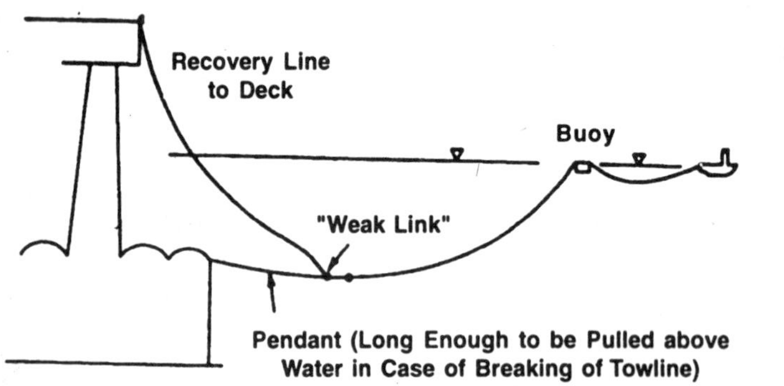


**Figure 5.11: Towline Arrangement for Deep-draft Tow**

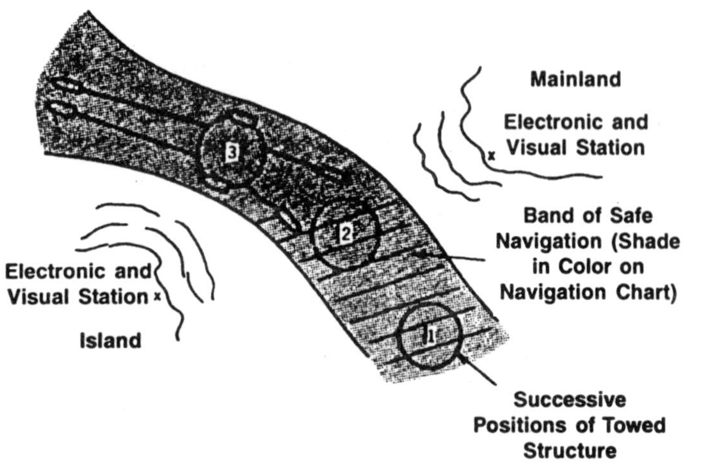


**Figure 5.12: Laying Out a Safe Swath or Band for Navigating in Constricted Channels**

Amazon, and Congo. In some cases, especially if crossing a bar, heave response may need to be considered.

The usual requirement for underkeel clearance is that the distance between maximum static draft and minimum water depth should not be less than 2 m or 10 per cent of the maximum static draft, whichever is lesser, plus an allowance for motion. The maximum static draft should be the actual measured draft at the deepest point with allowance for errors in measurement, initial trim, and water density change. The motion allowance should include the maximum increase in draft due to towline pull, wind heel, roll and pitch, heave and squat. These values can best be determined by model tests.

Air cushions may be used to reduce draft when crossing local areas of limited water depth. In general, the use of air cushions should be employed only to increase underkeel clearance above the theoretical minimum value to ensure the structure will still not hit in event of loss or air. It is important that the reduction in metacentric height and stability due to an air cushion be considered, since the air cushion acts like a free surface in reducing the moment of inertia. After the crossing of the shoal area, the air cushion should be completely vented. When using an air cushion, an adequate water seal height between the skirts must be left to prevent loss of air. The height of the seal will depend on the speed, since this will cause some air to be sucked out. Typical water seals vary from 0.5 to 2.0 m in height.

To optimize the reduction in draft, large air bags, each 11 × 11 × 4 m, of PVC-coated polyester fabric, were inserted under the base of the Andoc Dunlin platform. This enabled the full depth of the skirts to be utilized for buoyancy, without need for a water seal. Free air was also used in the compartments in the small spaces around the bags.

Communication between multiple boats during a critical positioning operation is all-important. Voice communication is used exclusively; however, it must be remembered that the tug skippers are of all nationalities. While English is usually the common language, lack of full comprehension and misunderstandings have led to serious mistakes. To obviate these, a carefully agreed set of common procedures should be adopted and reviewed so that there will be a clear understanding of all commands. If there are one or more captains who are not fluent in English, it may be desirable to have an interpreter available.

Procedures should be adopted to handle the case of a broken towline. The boat in question must take in the line and circle back. The towed vessel or structure should recover the bridle or pendant. As the towboat returns alongside, a messenger can be passed and the towline brought on board and made fast. All this is simple with one boat in a calm sea. It is very complex when it occurs at night in heavy seas and wind and the boat is one of three or four, each with its tow lines under strain. For very long tows, its provision must be made for refueling en route. One boat at a time can be fueled from a spare boat.

The dynamic accelerations of the towed structure should generally be limited to 0.2 g, to minimize forces acting on the tie-downs and to minimize adverse effects on personnel.

Emergencies which must be included in the planning for a tow are fire, flooding, and man overboard. While in congested areas near the exit port, a special fast boat "guardship" should be employed for the dual purpose of picking up a man overboard and warning away sight seeing boats. When the Condeep Statfjord A sailed from Stavanger, private cruise boats advertised an excursion alongside. This involved hundreds of people whose safety was paramount. News photographers fall into a similar category. Imagine what would happen if a tour boat was overturned by running up on a submerged towline! Harbor police can often be engaged to keep the route clear. In the case of the Statfjord A Condeep, the ceremonies were held on the advertised day, the flags flown, the cruise boats ran around to take pictures, and the towboats blew their whistles. Meanwhile the caisson was securely moored. Two days later, with no fanfare, the flotilla actually got underway, with no unwanted interference.

When large and valuable structures are towed (Statfjord B and C each had a value of approximately U.S. $2 billion), the insurance surveyors require a full manning, with adequate pumping capacity, power generation, and firefighting capabilities. Manning of large and important structures under tow may require a crew up to 30 in

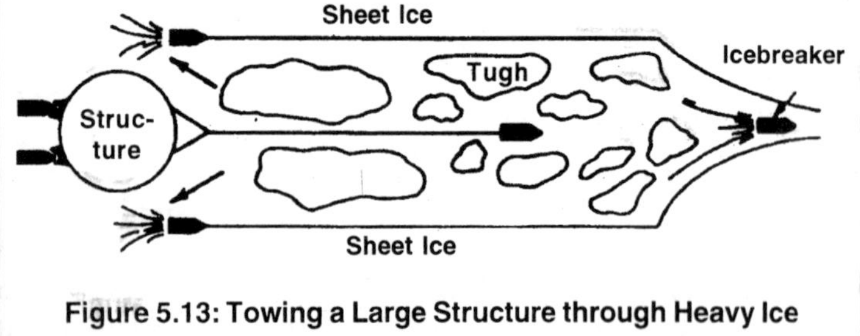


Figure 5.13: Towing a Large Structure through Heavy Ice

number. This crew will probably contain not only marine and ballasting crews, but also a meteorologist, sonar specialist, and navigating surveyors.

Some of the structures proposed for the Arctic have a conical shape. In the open sea, waves can run up over the lower sides, leading to erratic response. The waterline diminishes rapidly with immersion; hence stability can be significantly reduced. Tows of such structures may require trim down by the stern, thus increasing draft. In broken ice, masses may ride up over the sides. The effect of all these can best be evaluated by model tests.

Tows of lesser value may be manned or unmanned. If manned, the Coast Guard will require adequate life rafts and communications to ensure the safety of those on board. If unmanned, trailer lines of fiber should extend from the four corners or quarters to enable a boat to pick up the tow and put personnel on board.

When touring a deep-draft structure through shoot water, the tidal conditions must be carefully plotted. Delays of a few hours, common to marine operations, must be anticipated; otherwise the structure may arrive at the critical zone at low tide. On the other hand, an advantage of traversing a shoal at mid-tide is that, if the structure does ground, it will probably raise off at the next high tide. Of course, shifting ballast or deballasting may also be attempted, but not so much as to endanger stability once the structure floats free. Currents affecting the tow of large structures are principally tidal in nature. When towing the Ninian Central platform through the Minch, northeast of the Isle of Skye, the necessary height of high tide was unfortunately always preceded by an adverse flood tide. The current was such as to permit almost no headway to be made against the flood. Therefore the boats had to catch the slack water at the top of the tide and move over the shoal at maximum speed.

Summer storms may arise despite the best long-range and short-range forecasting. They may turn out to be so severe as to make it necessary to cut loose the tow. Lay-by and standby areas along the route should be identified and marked on the chart. The tow can proceed to point A; then, based on the current sea conditions and the short-range forecast, continue to B; and so forth. At each such station, the alternative of standing-by can be considered. These standby areas are selected for having adequate sea room to lee.

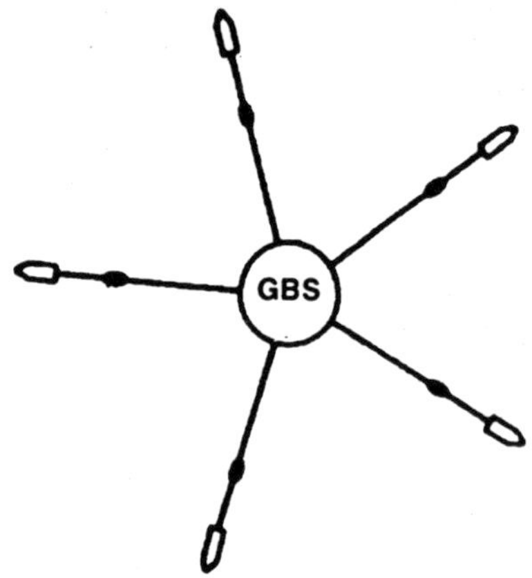


**Figure 5.14: Positioning of Tugs to Enable Accurate Installation of a Large Gravity-base Structure**

When positioning a structure at an offshore site, it is customary for the tugs to fan out in star fashion. Then the positioning is controlled by going ahead on some tugs more than others; that is, all lines are kept taut. Such an arrangement has been used on the North Sea offshore concrete platforms. Note the use of buoys in the lines to prevent pulling the sterns under. Bow thrusters are very desirable in enabling a boat to turn into the wind without exerting an increased pull on its towline.

As noted earlier, towboats are usually rated by their indicated horsepower (IHP), whereas a more meaningful figure is the bollard pull which they can exert. A tug will not, of course, be able to maintain its static bollard pull under continuous running conditions at sea, since the bollard pull decreases with speed.

The towing horsepower selected should be sufficient to hold the towed structure against waves of $H_5 = 5$ m, 40-knot sustained wind, and 1-knot current. Obviously, these arbitrary parameters have to be adjusted to the region involved.

## Diving and Underwater Activities

Diving for inspection purposes may employ a self-contained oxygen supply. For construction purposes, an umbilical cord is generally employed. If an air mixture, oxygen plus nitrogen, is used, the blood will absorb nitrogen under pressurization. Under subsequent decompression, if carried out too rapidly, the nitrogen will form bubbles in the bloodstream, leading to serious injury and even death. This is the well-known nitrogen narcosis or "bends." The use of carefully developed gas mixtures such as helium/oxygen has enabled safe diving to be carried out to depths of 150 to even 300 m. Other gas mixtures have been developed and used to permit operations in even deeper waters. Too rapid a rate of ascent, even from a shallow dive, will lead to the bends.

Decompression rate schedules have therefore been developed to ensure that the blood's natural balance of gases can be restored without the formation of bubbles. Foremost among these and widely adopted by regulatory bodies throughout the world are the U.S. Navy's Standard Tables: and "Decompression Tables for Standard, Exceptional, and Extreme Exposure" and "Single and Repetitive Dives, Helium/Oxygen Tables." A decompression tank must always be available on or near the site, depending on the depth of dive.

Saturation diving systems and "bounce techniques" have been developed which take advantage of the physiological fact that after the blood becomes largely saturated with an inert gas such as helium, then further absorption proceeds very slowly. This enables a person to stay at depths for long periods of time, working short periods, resting without decompression in a habitat or chamber, then returning for another stint of work. Using deck decompression chambers, personnel transfer capsules, and a habitat, the saturation diving system can support several teams of divers for long periods of time. With bounce techniques, the divers periodically descend to work, then return to partial decompression only.

The time required to perform work underwater increases with depth, although not as rapidly as one might anticipate. For example, a number of tests have shown that specific items of work required 20 per cent longer when at shallow depths and 50 per cent longer at deep depths, as compared with the time required for performance in the air. The greatest components of time are the times for descent, ascent, and decompression. The rate of descent is usually important only for deeper dives. Decompression tanks are carried on deck of the diving support vessel to enable saturation diving to be employed and to repressurize a diver who develops symptoms of the bends.

A major limitation when using divers for underwater work is that of developing a reactive force, "getting a foothold" so that the diver may exert a force. Because the diver is in a state of near-neutral buoyancy, the diver is like a person in space; the diver shoving against a pipe, merely moves the diver away. When a diver can plant the feet firmly on the bottom or against a structure, the diver can exert 100 to 300 N (22 to 66 lb) force in push or pull. Various means of attachment and hydraulic tools have therefore been developed to assist the diver in exerting a force.

There are many forms of diving, based on the equipment used. These vary from the scuba gear to the new diving suits that protect the diver from injury and light-weight helmets with improved vision capability. Hard hats and full suits are required for underwater construction work, where the diver must be protected from abrasion and puncture as well as from debris. Wet suits are most commonly used. For deep diving and diving in the Arctic or sub-Arctic, the suits are heated, usually by warm water circulation. Free-swimming (scuba) divers can perform inspection tasks; the are limited in communication and, of course, have no power supply for tools. The concern is for potential injury to the lightly protected scuba diver, from jagged protrusions. Conversely, the scuba diver is very mobile and can quickly report on conditions, especially in areas of good visibility. Scuba diving is severely limited in a current greater than 0.4 knots (0.2 m/s).

Tethered divers can have warm water circulation for their suits, hard wire or fiberoptic links for communication, and hydraulic power. To enhance the diver's capabilities for work, diving chambers or bells may be used. These give the diver more freedom from encumbrances but, of course, limit mobility. Diving bells, operating at atmospheric pressure enable inspection and work by engineers who are not qualified divers. A complete pressure-resistant diving suit, known as "Jim" has been developed, which enables the diver to stay at atmospheric pressure. A refined version, "WASP,"' enables a worker to descend to 600 m depth. These are very bulky which may cause problems of entry and maneuverability in confined spaces.

The current trend in manned diving systems appears to be toward diving bells and similar systems to enable better and safer control. However, there are many tasks requiring entry into congested

**Table 5.1: Manned Diving Systems Employed in Offshore Construction**

| *Type of System* | *Working Depth (m)* | *Endurance* | *Cost Ratio* |
|---|---|---|---|
| Scuba (air) | 0–40 | Very short; interrupted ascent | 1 |
| Scuba (air) | 40–70 | Very short; decompression required | 2 |
| Helmet (air) | 0–70 | Limited by diver's physical endurance | 3 |
| Helmet (helium/oxygen gas) | 50–100 | Limited by diver's physical endurance | 4 |
| Bounce (2 divers) | 70–100 | Few days | 10 |
| Bounce (4 divers) | 70–100 | Over 10 days | 13 |
| Saturation (4 divers) | 70–300+ | Unlimited | 16 |
| Manned submersible (without diver lock-out) | 600–1500 | Moderate | 15 |
| Manned submersible (with diver lock-out) | 70–200 | Unlimited | 25 |
| 1-Atmosphere gear (*e.g.*, JIM, WASP) | 600 | Limited by diver's physical endurance | 6 |
| 1-Atmosphere diving bell | 1000 | Unlimited | 5 |

spaces and among congested bracing, which can only be done by a diver.

Diver communications is an area in which there have been major advances in recent years. In addition to hard-wired and fiber optic systems, modulated sonar-frequency carrier systems give ranges of 150 to 500 m and single-sideband communications can give a 1000 to 1500 m range.

One of the most serious limitations of diver work is inability to determine one's position. This is due to lack of visibility, to disorientation, and to the lack of reference points. Consider an underwater concrete caisson that to the diver presents an endless wall 60-m high. Markings are required. Large orange epoxy numerals have been painted at about 10-m spacings to assist divers in determining their location. Wire guide lines have been stretched to serve as guides for divers and to hold their position in a current. Such wire guide lines are especially important if a diver must enter under or through structure–for example, into the middle of a braced jacket–or underneath a gravity-base structure while it is still in the floating mode during construction, or into an outfall sewer. One problem of divers is that of marking locations so they can return. In addition to a wire line, accoustic pinger locators are often used. The diver may use a handheld sonar to enable a search for a pipeline or dropped object.

To clean off marine growth for inspection, high-pressure water jets as well as hydraulically operated rotary brushes have been developed. Both sonic and tethered guide line systems are employed to guide divers working under ice, so that they may safely return to the entry/exit hole.

Hard hat divers require extra weight and a taut wire line tether to descend in a current greater than 1.5 knots (0.8 m/s). Taut wire line tethers increase both safety and efficiency, enabling the diver to descend to a specific location, with both hands free to work. Rings and/or raised markers preaffixed to the structure enable the diver to move progressively and with proper orientation, even when turbidity completely prevents vision.

When bolts or turnbuckles or other heavy objects are to be connected underwater, they should be preattached to one structural element. Then the diver can raise them (or direct their raising by a line from the surface) and make the connection. Raising is safer than lowering, since lowering may accidentally hit the diver. For entry into pipelines or under structures or through bracing, a second diver should be used to tend the first diver's lines.

For especially complex tasks, a full-scale mock-up can be constructed on shore, so that divers may not only practice but, by blanking out their vision, learn the feel of every element. This was carried out very successfully for installing the complex temperature control fixtures on the upstream face of Shasta Dam at a depth of 100 m.

A major problem in diver communications is that of transmitting the information to the surface in a form that is fully understood. The distorted "duck-like" talk of a diver on helium/oxygen gas is well known. In addition to transmission of voice communications and data transmission by fiber optics, video has become a major means. The ability for an observer on deck to see what the diver sees in real time represents a tremendous advance. Underwater photography can, of course, be used for recording more clearly specific objects such as a welded joint. Both video and photography require a powerful light source; much recent development has been directed toward appropriate frequencies to reduce scattering and incident angle refraction distortion. In areas of limited visibility, diver-held videos have proved more successful than ROVs.

Many tools and procedures have been developed to enable divers to work effectively underwater. Among these are the following:

1. Wet welding techniques, using a high-velocity jet of inert gas to create a water-free zone. Wet welding can be used on low-carbon steels to as deep as 70 m. Although its qualities are strongly influenced by depth, hence pressure, satisfactory welds have recently been completed at 110 m.
2. Dry welding, using a habitat and employing gas metal and gas-tungsten arc techniques. Hyperbaric welding has been carried out to depths of over 1000 m.
3. Underwater cutting using the electric arc method. With a skilled diver, steel can be cut almost as rapidly as above water. Arc-flame methods can be used to depths up to 2000 m. At greater depths, potential problems exist as the density of water and gas tend to equalize due to the high pressure. Arcair has developed an electrode which will cut both concrete and steel.

4. Mechanical casing cutters and abrasive jet cutters.
5. A wide variety of hydraulically driven velocity power (explosively driven) tools have been developed in recent years, many of them by the Naval Civil Engineering Laboratory at Port Hueneme, California. These include actuators, impact wrenches, rotary brushes, rock drills, thermic lancers capable of cutting steel and concrete and even rock, explosive (power-actuated) pin-driving tools, grout dispensers (for epoxy injection), and NDT inspection devices. In addition to the use of conventional hydraulic fluids, seawater power supply systems have been developed.

Diving and divers are really a transportation system to enable work to be carried out in an otherwise inaccessible environment. Because of the inherent limitations that still exist because and of the high costs, experienced constructors make every effort to eliminate or reduce diving requirements. For those still required, extensive planning is devoted to the diver's support, transfer, and work conditions to maximize his safety and efficiency.

Now let us try to learn few more details about personnel transfer at sea.

The transfer of personnel from crew boat to offshore derrick barge or onto a fixed platform is a critical operation from the point of view of both safety an efficiency. In fact, the ability to move personnel on and off can become the limiting criterion for continued operations in a rough sea. All too often this operation is overlooked in the planning phase. In some bridge work, shallow water near the bridge ends may prevent the use of crew boats, especially at low tide.

The boat in which the personnel are travelling to the offshore rig is responding to the wave action in all modes, heave, pitch, and roll being the most critical for the transfer operation.

More common, therefore, are other means of transfer. It must be also considered that there is usually a substantial height differential between boat deck and barge or platform. Properly fendered boats can come alongside a large derrick barge under favorable sea conditions, using the barge as a breakwater. In the case of head seas, they may come up to the stern. A notch should be fabricated so that

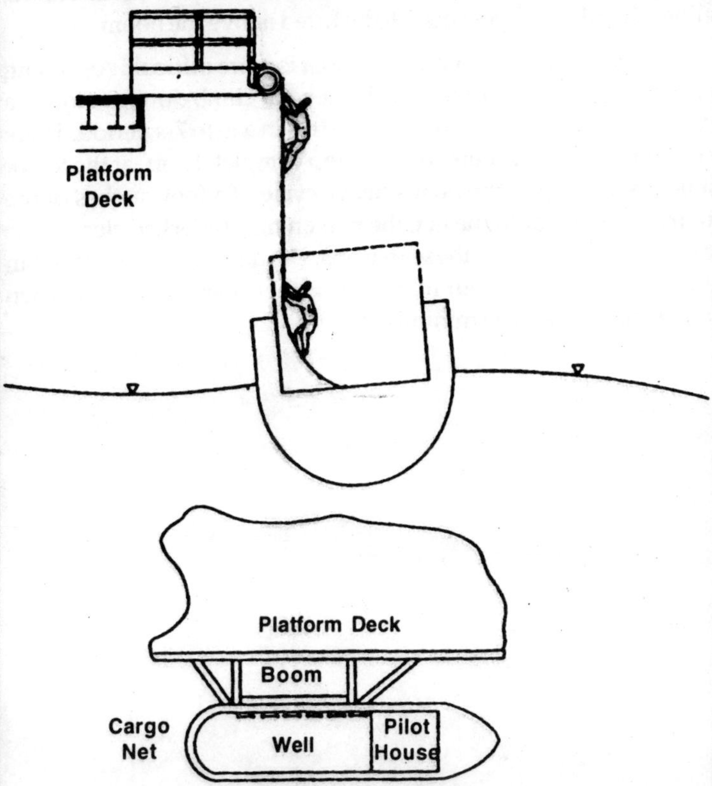


**Figure 5.15: Arrangement of Boom and Cargo Net for Personnel Transfer**

the boat can push tightly against the barge and thus minimize pitch response.

The cargo net concept has been adopted from the World War II troop transport experiences. It is hung from a boom, so that the lower end is at sea level; when the boat moors, the net can be hauled into the boat. It is a relatively simple and safe operation for people to catch the top of the heave–pitch cycle and climb up the net. When they reach the boom, however, they face a dilemma. Somehow they are expected to scramble onto the boom and walk to the deck. Even if the net is hung directly over the platform side, it is very difficult to scramble onto the deck unless handholds are provided. Hence, lifelines (handhold lines) need to be fitted above the boom.

Transfer back from platform to boat is more difficult. Assuming lifelines make it easy to get onto the net and climb down, below is a boat moving up and down several meters in a 5- to 7-s period. There is an instant when the person must step completely off, at the top or just before the top of the boat's heave cycle. If a foot catches or the man tries to hang onto the net, the person may be jerked clear of the boat as it descends. For these reasons, the net should be placed in the well of the boat, about midships, rather than at the bow. Then relative motions will be minimized.

## Chapter 6

# The Offshore Electrical Power

### Introduction

The provision of electrical power offshore involves practices similar to those likely to be adopted in onshore chemical plants and oil refineries.

The purpose of any offshore electrical supply system is to generate and distribute electricity to the user such that:

1. Power is available continuously at all times that the users equipment is required to operate.
2. The supply parameters are always within the range that the user's equipment can tolerate without damage, increased maintenance or loss of performance.
3. The cost per kilowatt hour (kWh) is not excessive, taking into consideration the logistical and environmental conditions in which generation and distribution are being effected.
4. Impracticable demands are not made on the particular offshore infrastructure, *i.e.*, such as those for fuel or cooling medium.
5. The safety requirements pertinent to an offshore oil installation are complied with, in particular those associated with fire and explosion hazards.

With the obvious availability of hydrocarbon gas as a fuel, and the requirement for a high power-to-weight ratio to keep structural scantlings to a minimum, gas turbines are the ideal prime movers for power requirements in excess of 1 MW. Below this value, reliability and other considerations tend to make gas turbines less attractive to the system designer.

Owing to the complexity and relative bulk of gas turbine intake and exhaust systems, the designer is urged towards a small number of large machines. However, he is constrained by the need for continuity of supply, maintenance and the reliability of the selected generator set to an optimum number of around three machines.

A variety of voltages and frequencies may be generated, from the American derived 13.8 kV and 4.16 kV 60 Hz to the British 11 kV, 6.6 kV and 3.3 kV 50 Hz. Many ships operate at 60 Hz, including all NATO warships, and there is a definite benefit to be gained from the better efficiencies of pumps and fans running at the 20 per cent higher speeds.

On most platforms, smaller generators are provided to maintain platform power for services other than production. These are also normally gas turbine driven and can provide a useful blackstart capability, especially if this is not available for the main machines.

The design of the distribution configuration at the platform topsides conceptual stage is very dependent on the type of oil field being operated and the economic and environmental constraints placed on the oil company at the time. The older platforms originally had few or no facilities for gas export or reinjection, and therefore the additional process modules installed when these facilities were required have their own dedicated high-voltage switchboards. This is also the case if the power requirement for such a heavy consumer as sea water injection is underestimated at the time of construction.

In general, however, it is better to concentrate switch rooms in one area of the platform in order to avoid complications with hazardous areas, ventilation etc.

With such relatively high generation capacities and heavy power users within the limited confines of an offshore platform, calculated prospective fault currents are often close to or beyond the short-circuit capabilities of the MV switchgear designs available at the platform topsides design phase. Currently, fault ratings of 1000 MVA are available, and with careful study of generator decrement curves etc. it is usually possible to overcome the problem without resorting to costly and heavy reactors.

All the available types of MV switchgear are in use offshore. The use of bulk oil types, however, is questionable owing to the greater inherent fire risk.

Unlike land based switchboards, there has been found to be a significant risk of earth faults occurring on the busbars of offshore switchboards, and so some form of earth fault protection should be included for this.

The platform distribution at medium voltage normally consists of transformer feeders plus motor circuit breaker or contactor feeders for main oil line (MOL) pumps, sea water lift and water injection pumps, and gas export and reinjection compressors. Depending on process cooling requirements, cooling medium pumps may also be driven by medium-voltage motors.

Operating such large motors on an offshore structure (*i.e.*, on the top of a high steel or concrete tower) can lead to peculiar forms of failure owing to the associated vibration and mechanical shock, almost unheard-of with machines securely concreted to the ground. This has led to offshore platform machines being fitted with more sophisticated condition monitoring than is usually found on similar machines onshore.

Another problem, is the transient effect on the output voltage and frequency of the platform generators with such large motors in the event of a motor fault, or for that matter during the normal large-motor switching operations. Computer simulation of the system must be carried out to ensure stability at such times, both at initial design and when any additional large motor is installed. Facilities such as fast load shedding and automatic load sharing may be installed to improve stability and also make the operator's task easier.

Using conventional oil or resin filled transformers, power is fed to the low-voltage switchboards via flame retardant plastic insulated cables.

But trunking is often used for incoming low-voltage supplies from transformers. Owing to competition for space, this is just as likely to be due to bending radius as to current rating limitations of cables, since bus ducting may have right angle bends.

The type of motor control centre switchboard used offshore would be very familiar to the onshore engineer. However, the configuration of the low-voltage distribution system, to ensure that alternative paths of supply are always available, is usually much more important offshore. This is because, although every effort is put into keeping it to a minimum, there is much more interdependence between systems offshore.

A few examples of small low-voltage supplies which are vital to the safe and continuous operation of the installation are as follows:

1. Safe area pressurization fans.
2. Hazardous area pressurization fans.
3. Generator auxiliaries.
4. Large-pump auxiliaries.
5. Large-compressor auxiliaries.
6. Galley and sanitation utilities for personnel accommodation.
7. Uninterruptible power supply (UPS) systems for process control and fire and gas monitoring.
8. Sea water ballast systems on tension leg platforms and semisubmersibles.

On any platform, there are a large number of systems which require supplies derived from batteries to minimize the risk of system outage due to supply failure.

The following is a typical platform inventory:

1. Fire and gas monitoring and protection.
2. Process instrumentation and control.
3. Emergency shutdown system.
4. Emergency auxiliaries for main generator prime movers.
5. Emergency auxiliaries for large compressors and pumps.
6. Navigational lanterns and fog warning system.
7. Emergency and escape lighting.
8. Tropospheric scatter link.
9. Line of sight links.

10. Carrier multiplexing and VFT equipment.
11. Telecommunications control and supervisory system.
12. Public address.
13. General alarm system.
14. Platform PABX.
15. Marine radio telephones.
16. Aeronautical VHF (AM) radio.
17. VHF (FM) marine radio.
18. Aeronautical non-directional beacon.
19. Company HF ISB and UHF (FM) private channel radios.
20. Telemetry system.
21. Satellite subsea well control systems.

The majority of these systems operate at a normal voltage of 24 V DC and, although it is not necessary for each of the above systems to have separate batter and battery charger systems, the grouping criteria require more detailed discussion.

In addition to the above systems there are, of course, switchgear tripping/closing supplies and engine start batteries which are dedicated to the equipment they supply. In the case of engines which drive fire pumps, duplicate chargers and batteries are also required.

## The Environment on an Offshore Installation

The environment on an offshore installation is not inherently safe, owing to the heavily salt-laden atmosphere and the highly conducting nature of the structure and virtually all the equipment it contains.

It must not be possible for personnel to come into contact with live or moving parts either by accident or while performing their normal duties.

Protection against electrical shock relies on the safe design and installation of equipment, on training personnel to be aware of the dangers and to take the necessary precautions, and on the use of special safe supplies for most portable equipment.

An electric current of only a few milliamperes flowing through the human body can cause muscular contractions and, in some

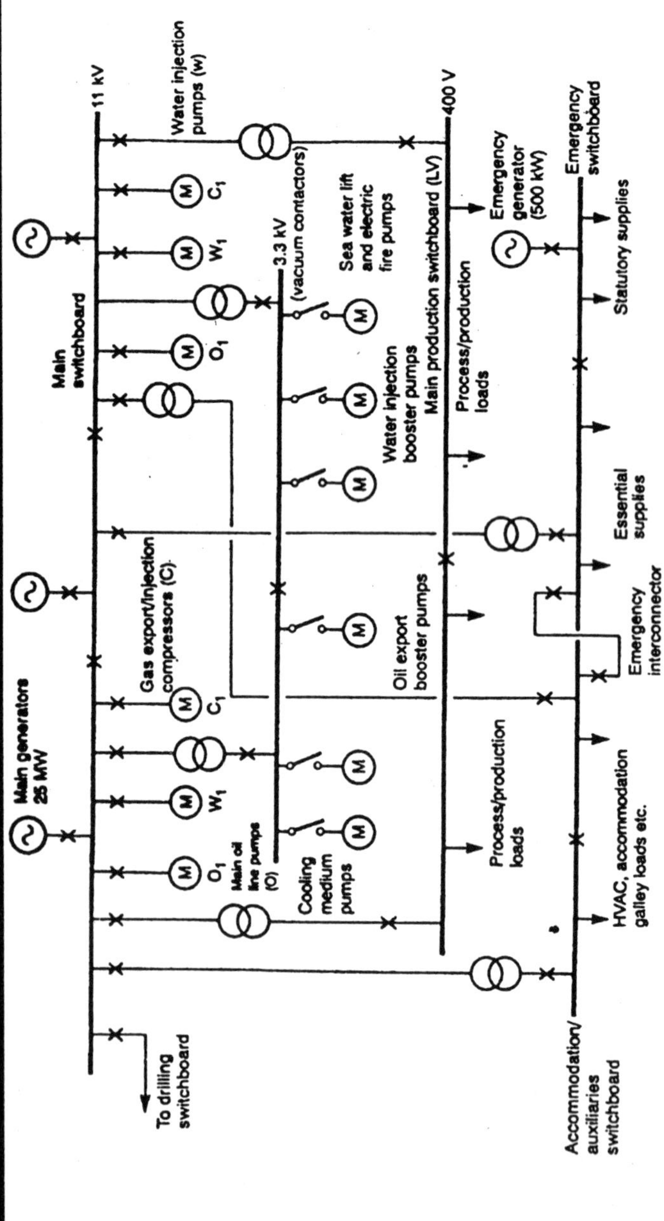


**Figure 6.1: Single-line Diagram of Typical Offshore Electrical System**

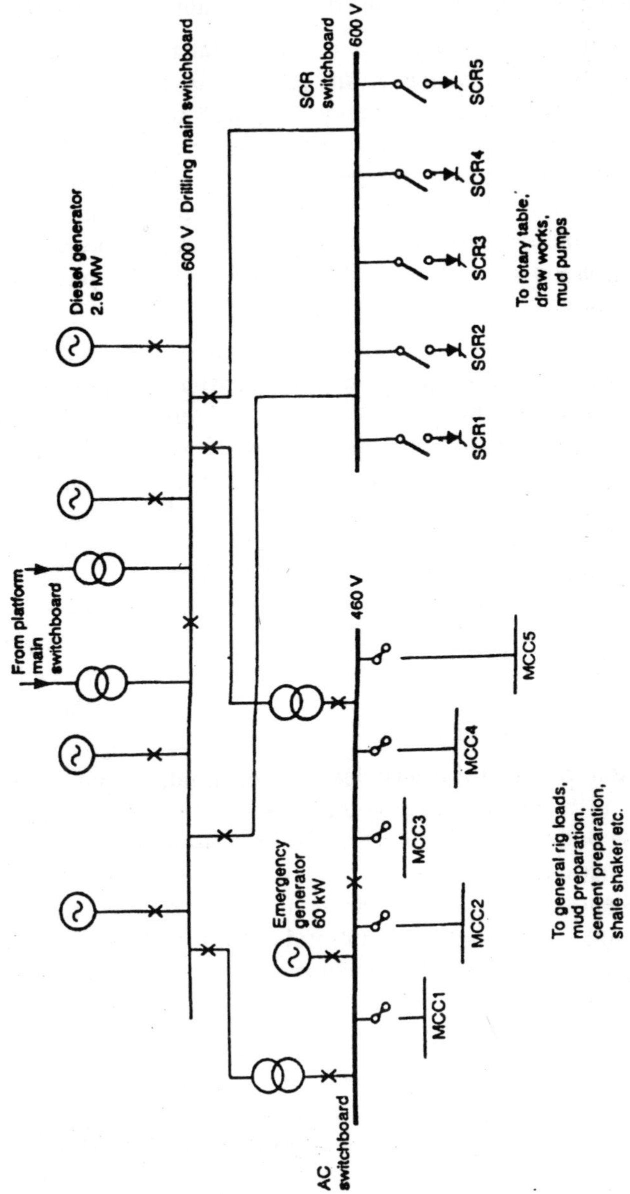


Figure 6.2: Single-line Diagram of Typical Drilling Electrical System

circumstances, will be fatal. The current may result in local burning or some involuntary reaction which in itself may lead to injury. Additionally, of course, varying degrees of pain will be experienced.

Except in the case of one or two installations, the electrical system is totally isolated from any other means of electrical supply. The system must be designed and configured in such a way that it is never dependent on one small component or electrical connection to continue in operation. This point may sound rather obvious, but it is the author's experience that hidden vulnerabilities may be designed into systems which are both costly and disruptive in their first effects and in their eradication. The following examples of actual occurrences illustrate the point.

A platform has two low-voltage switchboards dedicated to providing the safe and hazardous ventilation necessary for continued safe operation of the platform. Depressurization of any module would lead to a process shutdown. The particular platform is a pumping station for oil from other platforms, including those of other companies, and therefore considerable oil revenue is at stake if the platform is shutdown. Unfortunately each switchboard is fed by a single incomer, and the ventilation fan motor starters are distributed so that the majority of supply fans are on one switchboard and the majority of extract fans on the other. This arrangement resulted in the export of oil from a number of large North Sea installations being dependent on the continuous operation of two small low-voltage switchboards.

It is important to be very clear as to the fundamental reasoning behind all the regulations governing electrical installation offshore. Because both the safety and the cost of an installation are highly sensitive to equipment selection, it is also important to have a clear understanding of the reasons behind the classification of hazardous areas and of the different methods employed by equipment manufacturers to make their equipment suitable for particular environments.

Where this is practicable, electrical equipment is best installed in an environmentally controlled room which is located in an area unclassified with respect to hydrocarbon gas ignition risk, is effectively sealed from the outside atmosphere, and is provided with a recirculating air conditioning system. Of course, this optimum scheme cannot be considered for equipment which:

1. has to be located outside (such as navigational aids);
2. has to be located under or near water (such as sea water lift pump motors); or
3. is associated with some other equipment which may occasionally or does normally leak hydrocarbon gases (such as gas compressor drive motors).

Often the equipment installed has to safely cater for a combination of all three situations, and may also be required to operate at elevated pressures and temperatures.

Now let us try to understand about some details about hydrocarbon hazards.

In the planning of platform superstructures, designers try to arrange to segregate the wellhead and process areas from the accommodation and other normally manned areas to the greatest possible extent. This involves not only horizontal and vertical segregation but also segregation of all piped or ducted services such as ventilation ducting and drains.

Following the Piper Alpha disaster, it is likely that the whole philosophy regarding the segregation of accommodation areas on offshore platforms will be rethought. As is common knowledge, 165 men lost their lives either as a result of the initial explosions, dense smoke and fire, or following the ensuing riser fires which led to the loss of structural integrity and the falling of the accommodation modules into the sea. The recently published Cullen Report gives over 100 recommendations, covering all aspects of offshore installation design, construction, operation and safety. In one of the most important recommendations, Lord Cullen states that the operator should be required by regulation to submit to the regulatory body a safety case in respect of each of its installations. It is important to consider the safety aspects of each installation uniquely so as to meet objectives, rather than to impose fixed solutions which may or may not work on a particular installation.

Whatever further means of ensuring the survival of the particular installation and its personnel are considered in the safety case, it is certain to influence the design of the electrical system and equipment, particularly in minimizing the risk of electrical ignition sources and in the provision of emergency secure electrical supplies completely independent of normal platform supplies.

An important means of minimizing the risk of ignition sources are the hazardous area boundary drawings produced during the platform process design stage, which represent the situation during normal operating conditions. However, it is also necessary to consider the situation during a major outbreak of fire or after a serious gas leak–the so-called 'post-red' situation.

There are three systems which normally monitor and control the extent of oil and gas leaks and hence the safety of the platform:

1. The fire and gas monitoring system.
2. The emergency shutdown system.
3. Safe and hazardous area ventilation systems.

On floating installations, the ballast control system could also be included as a fourth in certain circumstances.

All these systems will have some bearing on the design of the platform electrical system, either because they may include the facility to shutdown all or part of the electrical system, or because a secure (or at least a more reliable) electrical supply is needed to operate them.

## The Platform Electrical Power System

The continuing safe condition of the platform cannot be left solely to the human operators, since they would not always have sufficient time to investigate each abnormality and respond with the appropriate sequence of corrective actions in every case.

Every installation has startup and shutdown systems of varying sophistication, which attempt to provide the greatest possible safety for personnel and equipment. These systems are interrelated, with the process control system being subordinate to the emergency shutdown system.

Most of the heating, ventilating and air conditioning (HVAC) systems must run continuously during normal platform operation to ensure that:

1. Acceptable working environments are maintained in process modules containing equipment or pipework which may leak hydrocarbon gas.
2. Comfortable environmental conditions are maintained within the accommodation modules and normally manned

non-hazardous areas, and an acceptable working environment is provided in normally unmanned modules.

3. Positive pressurization with respect to adjacent hazardous areas or the outside atmosphere is maintained in non-hazardous modules or rooms.
4. Potentially hazardous concentrations of explosive gas mixtures are diluted in, or removed from, hazardous area modules.
5. Individual areas are sealed from ventilation and the associated fans are shutdown in accordance with the logic of the emergency shutdown system, when fires occur or dangerous concentrations of gas are detected.
6. Uncontaminated combustion, purging and normal ventilation air is available to prime movers.
7. Uncontaminated air supplies are available to personnel, and emergency generator and fire pump prime movers and other essential service equipment are provided with combustion and ventilation air in times of emergency.

The three systems–fire and gas, emergency shutdown, and HVAC–are interconnected and are often required to work in concert. An example of this would be if a fire occurred in a particular switch-room. The fire would be detected by smoke or heat detectors, and the central fire and gas system monitoring the room would initiate the following actions:

1. Signal the ventilation system to seal the room by the closure of ventilation fire dampers and switch off associated fans.
2. Sound an alarm in the switchroom to warn personnel that escape is necessary and that a fire extinguishant is to be released.
3. Depending on the system logic, signal the emergency shutdown system to isolate the switchboards in the switchroom by opening the appropriate feeder circuit breakers, or even shutdown all main generators if the switchroom in question contains the main switchboard.
4. Release the fire extinguishant ($CO_2$ or halon gas) into the switchroom after a suitable time delay to allow for personnel to escape.

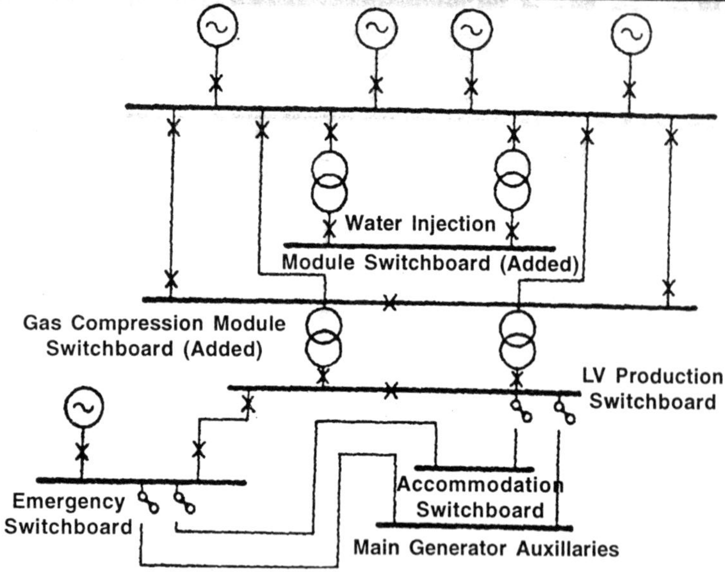


**Figure 6.3: Single-line Diagram of Early Installation with Added Utilities Switchboard**

Provided lifting and transport facilities of sufficient capacity are available at an economically viable cost, it is invariably better to build a complete module containing the generators, switchgear and transformers, completely fitted out, tested and commissioned at a suitable fabrication yard, than it is to carry out any of the construction on the platform.

Apart from fuel, cooling air and combustion air, it is preferable to make the electric power module totally independent of the rest of the installation. This also means that it has integral engine starting facilities as well as engine auxiliaries which do not require external low-voltage electrical supplies. This is not always possible owing to the weight of the extra transformers required. The need for sea water cooling for alternator heat exchangers may be unavoidable, owing to the bulk of air cooled units. The optimum independence of the module has the added advantages of:

1. Minimal hookup requirements during offshore installation.

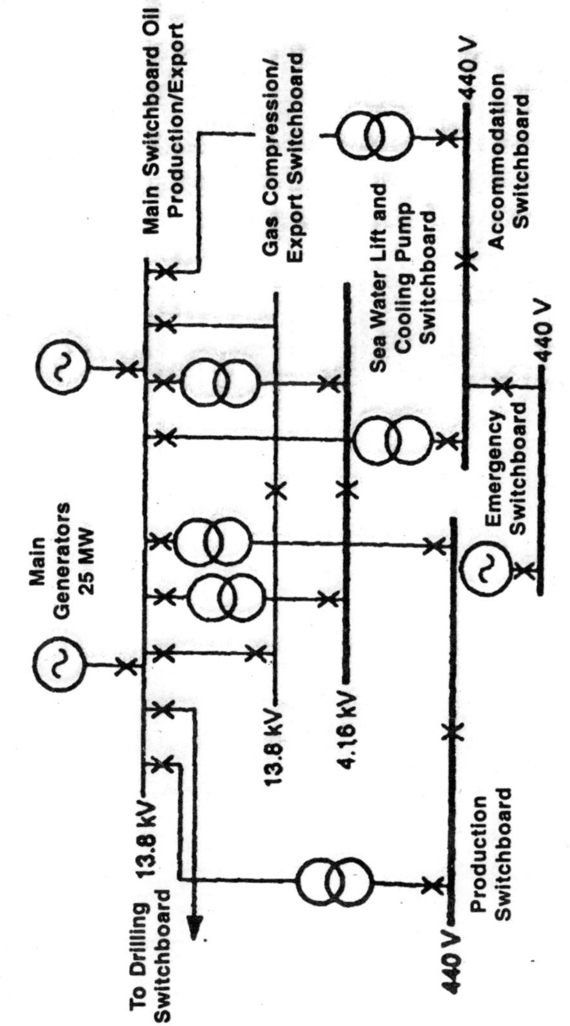


**Figure 6.4: Single-line Diagram of System with Two Platform Load Capacity Generators**

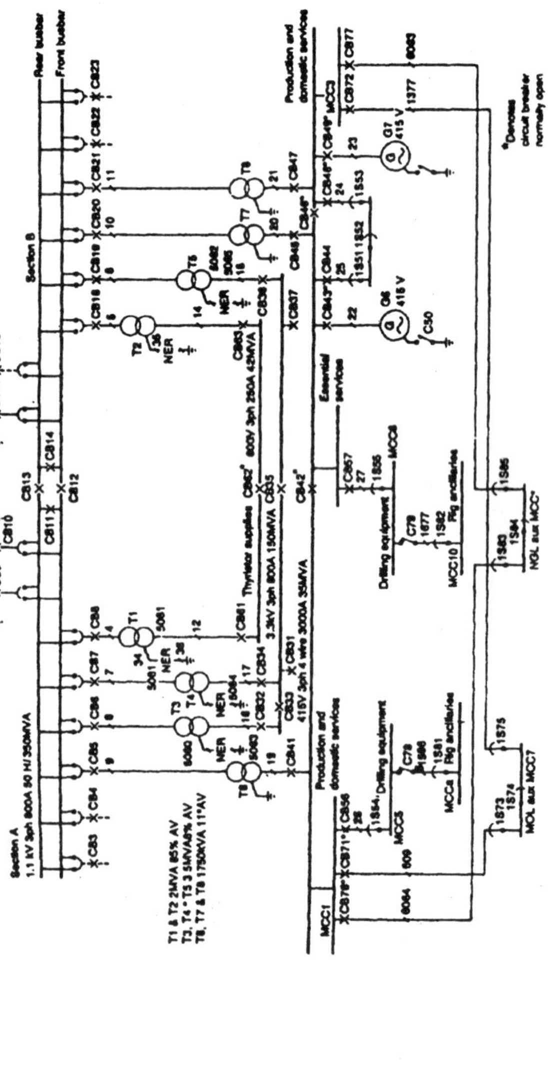

Figure 6.5: System with Three Main Generators Feeding Sectionalized Switchboard

2. Minimal service requirements during test and commissioning at the module fabrication yard.

If it becomes necessary to supplement the available power on a particular platform, then the additional weight of supplementary generators may be too great for the platform to bear without very costly modifications. Even if weight is not a problem, it is not always possible to find a sufficiently spacious location on the installation.

Shore supplies may be of the wrong frequency for use on the particular installation, and it may be necessary to install a motor-generator set. This has the additional advantage of improving the motor starting capability of the supply, as the generator impedance will be much lower than that of a series of transformers and long subsea cables.

The transmission voltage required will vary depending on the length of the subsea cable, but is likely to be either 11 or 33 kV. The weight and space taken up by the transmission transformers and the associated extra switchgear need to be taken into consideration whenever subsea cable options are proposed.

If there is a group of several small installations separated only by a few kilo metres of water, it may be economic to supply all their main power requirements from one central platform. This is more likely to be the case if centralizing the main generation allows gas turbines of 1 MW or more to be considered.

It is advisable to carry out some form of reliability analysis in order to numerically rank the reliabilities of various supply and generation schemes before making a final decision.

## Chapter 7

# The Offshore Piling

### Introduction

Piling for offshore structures must be installed to develop required capacities in bearing, uplift and lateral resistance.

Deep water, long, unsupported column lengths, large cyclic bending forces, and large lateral and axial forces all combine to make offshore piles large in diameter and long in length. Piles in most offshore practice are steel pipe piles ranging from 1 m up to 2 m (and even 4 m) in diameter and in lengths from 40 to 300 m. Marine piles for harbors are generally much larger and more heavily reinforced than piles for land foundations.

For resisting axial compression, the pile transfers its load by skin friction along its outside perimeter and by end bearing on its tip, provided that the tip is either closed or plugged in such a way as not to yield in relation to the pile. Thus for a natural plug of sandy clay, the internal skin friction must be adequate to develop the full end-bearing resistance of the plug. Large-diameter tubular piles may not plug, however, and thus the end bearing is lost. However, the interior surface will develop skin friction.

End bearing and skin friction do not develop their resistances simultaneously and hence are not usually directly additive at serviceability (elastic) levels of load. They may, however, partially augment each other at ultimate load. For this reason, deep pile foundations are usually designed primarily as friction piles. Internal skin friction generally develops to its maximum within a one diameter length of the tip.

For resisting axial tension, the deadweight of the pile plus that of the internal plug of soil plus the skin friction are available.

For resisting lateral loads, most offshore structures in deep water (over 30 to 40 m) depend on bending resistance of the pile interacting with the passive resistance of the soil in the near-surface stratum. Since the soil resistance is a function of its deformation, the analysis is based on the interaction of the lateral load $P$ with the displacement $y$ at each incremental level below the seafloor. Hence, this is called the $P/y$ effect. The pile must have sufficient strength to resist the resultant moment at these levels and to prevent local buckling. The capacity to resist lateral loads can be improved by increasing the stiffness and moment capacity of the pile in the critical zone near and just below the mudline by grouting in an insert pile with concrete, by increasing the wall thickness of the steel pile through this zone, and by filling the pile in this region.

In stiff clay soils and calcareous soils, cyclic lateral loadings may create a gap around the pile, just below the mudline, which increases the lateral deformations of the piles and structure as a whole and increases the moment in the piles. Piling a loose mound of pea gravel, or even high-density rock of small size, around the pile can effectively minimize this effect by filling any gap that does form and thus minimize the amplitude of deformations.

An alternative method of resisting lateral loads, used in harbor structures and some offshore terminals, is by the use of batter (raker) piles, sufficiently inclined to develop a substantial horizontal component of their axial capacities. Batter piles must have a reaction in order to be effective; this is usually provided by a mating pile battered in the opposite direction, although the deadweight of the platform may also be mobilized as a reaction force. Under overload, these raked piles develop substantial rotations; their performance in earthquakes and ship collision has generally been unsatisfactory due to local buckling near the pile head.

For offshore piling and for bridge piers in deep water and/or very soft mud, it is customary to use either all vertical piles or to have a moderate batter, up to 1 on 6. Under lateral forces and imposed deformations such as those of storm waves and earthquakes, the

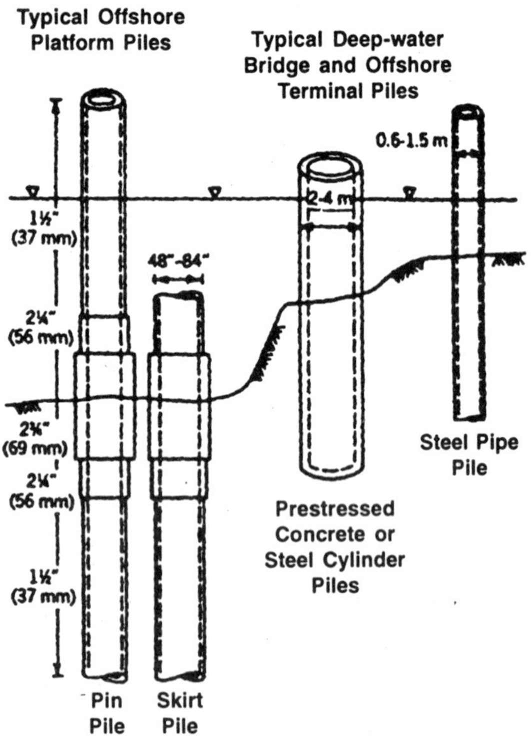


Figure 7.1: Typical Piles for Offshore Structures

increased bending moments and axial forces are generally within the capacity of the piles. The batter helps to eliminate any significant residual displacement of the structure.

Special methods and equipment have had to be developed to install the large piles required for offshore structures. Driving with very large hammers is still the preferred method for most cases because it is fast. However, where soil conditions do not permit driven installations and in other special cases, drilling may be employed, with the pile being grouted into the drilled hole.

The effect of all installation operations on the supporting soil must be considered. In some cases, it may be beneficial; in most cases, the results may degrade the performance unless special precautions are taken. API Standard RP2A warns that piles drilled and grouted may have resisting values significantly different from those of driven piles. Large-diameter piles (over 1.5 m) may not develop their full internal skin friction due to "friction fatigue."

For piles driven in undersized drilled or jetted holes in clays, the skin friction will depend on the amount of soil disturbance, including the relief of stress, which is occasioned by the installation. The strength of dry, compacted shale or serpentine may be greatly reduced when exposed to water from jetting or drilling. The sidewall of the hole may develop a layer of slaked mud or clay which will never regain the initial strength of the parent soil or rock.

In overconsolidated clays, drilled and grouted piles may develop increased skin friction. If excess drilling mud is present in clays or in soft rock, the coefficient of friction may be significantly reduced. In calcareous sands and some silts, driven piles may have very low values of friction compared to those attainable by drilling and grouting. Jetting seriously reduces the sheer strength of siltstone.

API Standard RP2A further warns that the lateral resistance of the soil near the surface is significant to pile design, and consideration must be given to the effects of soil disturbance during installation as well as scour in service.

## Driving of Offshore Piling

The piles for the typical offshore jacket are delivered on barges, with the first section of each pile being as long as can be handled and placed by the derrick barges. Pin piles are centered inside the jacket legs and typically extend the full height of the jacket. Skirt piles are encased in sleeves bracketed out from the lower end of the jacket. Many jackets incorporate both pin piles and skirt piles.

Skirt piles must be driven either with a follower of an underwater hammer. The piles are typically clustered around the corner legs of the jacket and are aligned parallel to them, so that the piles must be driven on a batter of from 1 to 6. Guides are attached at intervals along the jacket legs to aid in setting the piles through the sleeves.

Some recent jackets have been constructed with vertical sleeves, thus eliminating the guides and enabling a very long initial pile segment to be stabbed into the underwater sleeve. Guidance of the pile may be by means of a tensioned line or, in deep water, by the use of short-range sonar and video.

"Add-ons" (additional lengths of pile) are "stabbed" onto the top of the previous pile section as it is driven down near to the top of the jacket.

American Petroleum Institute Standard API RP2A suggests that reasonable assurance against failure of the pile will be provided if static stresses are calculated for each stage as follows:

1. The projecting section of the pile is considered as a free-standing column with a minimum effective length factor *K* of 2.
2. Bending moments and axial loads are calculated on the basis of the full weight of the hammer, cap, and leads, acting through the center of gravity of their combined masses, plus the weight of the pile add-on section, all with due consideration of the eccentricities due to pile batter. The bending moment so determined should not be less than that due to a load equal to 10 per cent of the combined weight of hammer, cap, and leads applied at the pile head perpendicular to the pile centerline.
3. The pile resistance is to be based on normal (elastic) stresses, with no increase for the temporary nature of the load.

One means of reducing the bending in the pile during this stage is to suspend the hammer and leads in a bridle at the proper batter. This is especially important in offshore terminal construction where relatively flat batters are often employed (*e.g.*, 1 horizontal to 2 vertical). This is also very important when driving piles having low bending strength–for example, prestressed concrete piles–on a batter.

Offshore pilings are typically large-diameter, thick-walled tubulars (pipe) ranging from 1 to 2 m in diameter. They are driven with high-energy impact hammers, either steam, hydraulic, or diesel. As a general rule, the hammer with attached driving head rides the pile rather than being supported by leads. This means that the driving head (helmet) must be secured to the hammer by wire rope slings and that the driving head in turn must seat well on the pile and have a guiding bracket or ring attached in order to keep the hammer aligned with the pile. During driving, the hammer line from the crane boom is slackened to prevent transmitting impact and vibration into the boom.

For steel piles, there is usually no cushion block used between the helmet and the pile, although an internal cushion is used in certain makes of pile hammer. Because of the tremendous energy required to raise the ram, steam or hydraulics are usually used rather than compressed air. Offshore hammers are generally single-acting with rates of up to 40 blows per minute. Hammer energies of current equipment range from 100 kN-m (67,000 ft-lb) to 1800 kN-m (1,200,000 ft-lb) per blow. The larger hammers represent major lifting tasks in themselves, weighing up to 300 tons.

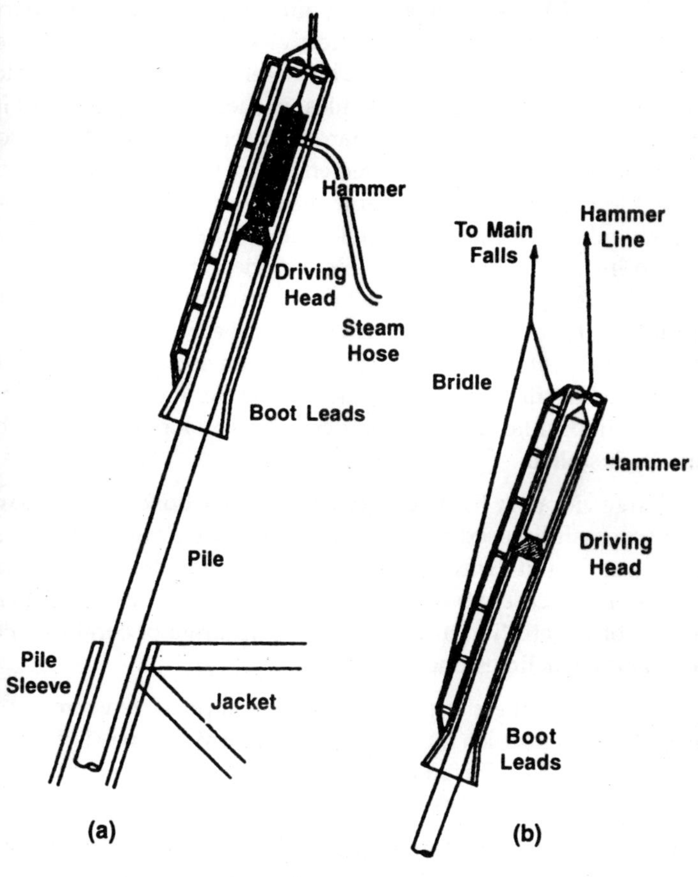


**Figure 7.2: Driving of Offshore Piling; Suspension of Boot Leads and Hammer**

Hydraulic hammers have been developed for offshore and especially for underwater driving. These are radically new versions of the underwater hammers formerly used in bridge pier construction. The new hammers not only have large energies, but are virtually unaffected by depth. They are thus useful for driving skirt piles whose heads may finally be located several hundred meters below sea level. Hydraulic hammers have a favorable action in that they sustain the impact over a longer number of milliseconds than a steam hammer.

Large diesel hammers are much used on offshore terminals. These have nominal energies in the range 200 kN-m (130,000 ft-lb) to 300 kN-m (200,000 ft-lb) per blow, but in most practicable driving conditions they can be equated in effectiveness with a steam hammer of 60 per cent of that rated energy. The diesel hammer is much lighter to handle and much more economical in fuel consumption, but its effective energy is limited. One manufacturer is developing a much larger diesel hammer, with a rated energy of 600 kN-m (400,000 ft-lb) per blow. Large vibratory hammers have been used on offshore terminal piling: for example, a quad unit of four large vibrators was used to install the heavy steel piles on the Yanbu, Saudi Arabia, pipe off loading pier. These actually activated the piles to a resonant amplitude of about 10 mm, and the pile "drove itself" through dense sands and limestone layers. Vibratory hammers have been used to initiate the installation of piles: they are lighter and shorter than a steam or hydraulic hammer and hence allow a longer length to be initially installed.

Large vibratory hammers, joined in dual or quad mounts, have been successfully used to drive steel piles including steel tubular piles up to 3 m in diameter, through moderately dense sands and gravels. In one case, a quad mount was able to drive 2-m piles to 20 to 30 m of penetration through alternating strata of coraline, silty sand, and weak limestone.

Impact hammers impart an intense compressive wave to the head of the pile, which travels down the pile at the speed of sound in the pile material. This compressive wave is a dynamic stress wave which eventually reaches the tip and extends it into the soil.

The newest large steam and hydraulic offshore hammers are instrumented so that the velocity of the ram can be measured just before it strikes the anvil. A representative pile can also be instrumented to measure both strains and acceleration during the hammer blow.

**Table 7.1: Large Diesel Hammers**

| *Hammer* | *Blows per Minute* | *Ram Weight (tons)* | *Total Weight (tons)* | *Energy (KN-m)* |
|---|---|---|---|---|
| Delmag D-200-42 | 36-52 | 20 | 50 | 680-436* |
| Kobe K-150 | 45-60 | 15 | 36 | 400 |
| Mitsubishi MB-70 | 38-60 | 8 | 21 | 200-90* |
| Delmag D-55 | 36-47 | 5 | 11 | 160-90 |
| Kobe K-60 | 42-60 | 6 | 17 | 145 |
| Delmag D 46-02 | 37-53 | 4 | 8 | 145-160 |
| Delmag D 65 | 37-53 | 8 | 10 | 165 |

* Adjustable.

Large-diameter tubular steel piles are being increasingly utilized, not only for offshore platforms but for marine terminals and bridge piers as well. To drive these piles requires high energies per blow, hence very large and costly hammers. Since the total number of piles in an individual project is low, an alternative pile hammer has recently been developed. This hammer is essentially a hydraulically-operated drop hammer. It consists of a long shaft, 5 m or more in length, filled with lead or concrete. The shaft is slightly smaller than the tubular which it will drive. The upper end is enlarged to include a driving ring which will impact on the tubular pile. This mandrel is raised and released by an external lifting frame, clamped to the pile. Hydraulically operated lugs extend and engage the mandrel. Hydraulic rams lift the mandrel 1.5 m. The lugs then release and the mandrel drops. Cycle times are 30 s to 1 min, thus driving times are long. This relatively low-cost drop hammer can develop very large impact energies. The mandrel must be free from stress concentrations; heat treatment after fabrication is needed. If this hammer is to be used underwater, then the mandrel must have a large central void to allow the water to escape rapidly. Such a hammer was used to drive a single 4-m-diameter tubular pile to 35-m penetration in a dense silt stone off the coast of San Diego, California, and is planned for use on a terminal in Puget Sound, Washington.

The $D/t$ ratios for piles must be limited to preclude local buckling at stresses up to the yield point of the pile steel. Where only moderate driving resistances are anticipated or where the pile will be drilled and grouted (not driven), the pile may be designed as a steel cylindrical member and checked for local buckling due to combined axial compression and bending. This latter is non-critical when $D/t$ is less than or equal to 60. When $D/t$ is greater than 60, then a more in-depth analysis such as that given in API RP2A should be followed.

For piles that will be subjected to sustained hard driving in excess of 800 blows/in. (250 blows/ft) the minimum wall thickness of the pile should be not less than:

$$t\,(\text{mm}) = 6.25 + D\,(\text{mm})/100$$

The head of pile sections on which driving is carried out may be deformed during driving and hence require reheading in order to weld on the next section. Hence API RP2A recommends an allowance of 0.5 to 2 m for reheading. Modern well-fitting driving heads and some hammers (*e.g.*, Hydroblock) minimize the head damage; so that with thick-walled piles, reheading may be unnecessary.

When pile add-ons are placed, they are equipped with stabbing guides to facilitate entry and proper alignment. The stabbing guides should have a tight fit in order to provide a proper fit-up for the weld. The guides may be designed to support the full weight of the pile during welding so that the crane may be freed from other tasks. Further, support from a floating derrick boom during welding is often unsatisfactory due to the movement of the boom tip and transmitted vibration. However, it is usual practice for the crane to continue to hold on with a slack pile line as a safety precaution until at least one full weld pass has been made. Pile sections may also be held by temporary supports on guides which extend up from the jacket and provide a support 10 to 20 m above the deck. They may also be held by a hydraulically operated clamping and alignment device which is clamped onto the previously set pile section or supported on a temporary work deck on the jacket. This latter allows quick-stabbing guidance and then final accurate alignment of the new section.

In addition to connecting pile sections by full penetration welds, breech-block connector sections have been developed which enable the splice to be effected rapidly by applying torque. Accurate alignment is essential, and hence a hydraulic clamping-aligning device is essential. These mechanical connectors have been used for both pile followers and permanent piles and have shown fully adequate performance during driving.

For piles to be joined by welding, the add-on section is pre-beveled, ready for a full-penetration weld. After stabbing, the bevel is inspected and, if necessary, ground or gouged to open it up to assure a full-penetration weld. Weld procedures and materials should be carefully selected with regard to the pile steel qualities and the temperature at which driving will be carried out, since these welds will certainly be under high impact. This will be especially critical for pile driving in Arctic and sub-Arctic regions which may be carried out at low temperatures. In any event, low-hydrogen electrodes should be used. Semiautomatic welding machines are becoming available in order to speed the welding process: these have 2 or 4 welding heads.

Backup plates are usually built into the stabbing guides. However, where drilling is to be carried out, internal backup plates cannot be employed, and the stabbing guides must be external to the pile. API RP2A notes that special skills are required for single-side welding or complete-penetration welds where backup pates are not used.

For long piles with numerous add-ons, great care should be taken to ensure accurate axial alignment of each section, so that the resultant full pile will be as straight as possible. This is especially important for clustered skirt piles. In order to keep their tips properly spaced.

The time required for splicing of the large-diameter, thick-walled piles typical of most offshore platforms is significant. On the 54 in. piles of the Hondo platform, the average times were for 1-in. walls, 3¼ hours; for 1¾ in. walls, 7¼ hours; for 2¼ in. walls, 10¼ hours. On the Jamuna Bridge in Bangladesh, for tubular piles of 2.5 and 3.15 m, with 50 to 60 mm wall thicknesses, four teams of welders were used simultaneously. Splicing took 6 to 8 hours, cool down 2 hours. X-ray 1 hour.

Welding machines should be properly grounded to prevent underwater corrosion damage due to stray current discharge.

When pile sections are lifted, they are usually provided with lifting eyes. The lifting eyes and their weld details are designed for the stresses developed both at initial pickup and as the pile is rotated to alignment with its final axis. Both angle of lead of the sling and the load acting on the padeye will change during this operation. An allowance must be made for impact during lifting- normally 100 per cent for the lifting eye and 35 per cent for the crane.

When lifting eyes or weld-on lugs are used to support the pile sections from the top of the jacket, each eye or lug should be designed to support the entire hanging weight. Lifting eyes for tubular pile sections must be welded transversely to the initial tensile force; there is no suitable alternative. Hence, it is especially important that a full-penetration weld (not fillet welds), with proper procedures, be used.

If the pile driving has been stopped for 1 hour or more in order to splice a pile or because of equipment malfunction, weather delays,

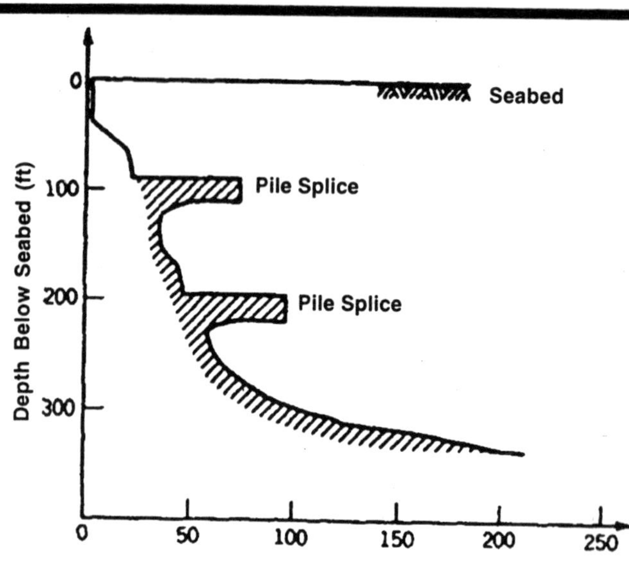


Figure 7.3: Typical Effects of Suspension of Driving During Splicing of Pile in Cohesive Soils

or the like, then the pile should be driven at least 0.3 m before the above criteria are reinstated. Driving at resistances greater than 800 blows for 150 mm should not be attempted. The pile will be deforming (yielding), and no appreciable penetration will be attained.

Approximate guide lines for selecting hammer size in relation to pile diameter and wall thickness are given in API RP2A. These do not include other important parameters such as pile length and soil characteristics, but they do address the problem of preventing excessive local damage in the pile due to dynamic stresses induced by the hammer and have been determined largely from industry experience in driving offshore piling in the medium to large sizes with moderate-sized hammers.

When steel piles are driven onto rock, the rebound compressive stress at the tip may be almost twice that imparted directly to the pile by the hammer, thus often being above yield. This condition is most frequently encountered in constructing offshore terminals. The pile tip may deform, tear, or "accordion." Tip reinforcement is definitely beneficial. The wave equation is a useful tool for prediction of pile tip stresses in such cases.

The trend today is toward the use of thicker pile walls in order to increase the effectiveness of the hammer in obtaining penetration. Heavier hammers are used in order to achieve more effective penetration and driving rates. A somewhat more conservative approach is therefore suggested.

When entering an initial section of pile into a sleeve, an extra long length may be stabbed, limited only by the bending moment due to pile deadweight alone and the lifting capacity of the crane. Once entered and run down, an axial force may be applied to run the pile farther into the soil. This is done by rigging a line around the head of the pile section, down through a snatch block affixed to the jacket near the pile sleeve, and thence to a winch on the jacket or derrick barge. Exertion of a tension on the line pulls that pile down through the soil to a temporary top elevation which is safe for mounting of the hammer. Similarly, on offshore terminal construction, with prestressed concrete piles, by simultaneously jetting at the pile tip (through an internal jet) and exerting a downward tension axially aligned with the pile, the pile may be caused to penetrate through soft clays and sands to a more workable elevation.

With large-diameter cylinder piles, the pile will have to be slung from padeyes on both sides of the pile (opposite ends of a diameter) to hang vertically.

Upending large cylinder piles presents a problem since it may be practicable to lift only from the top, yet there may be insufficient bending strength to permit such support. If the pile is on a barge, then a rail-mounted bogey car may support the lower end, yet move along the barge with the pile as it is raised to vertical. Another

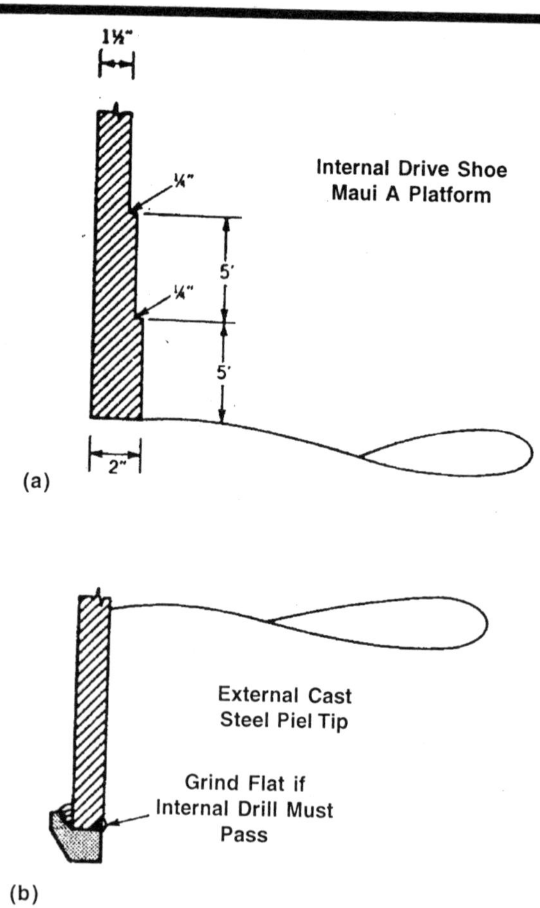


Figure 7.4: Tip Reinforcement for Piles Expected to Encounter Rock or Hard Layers

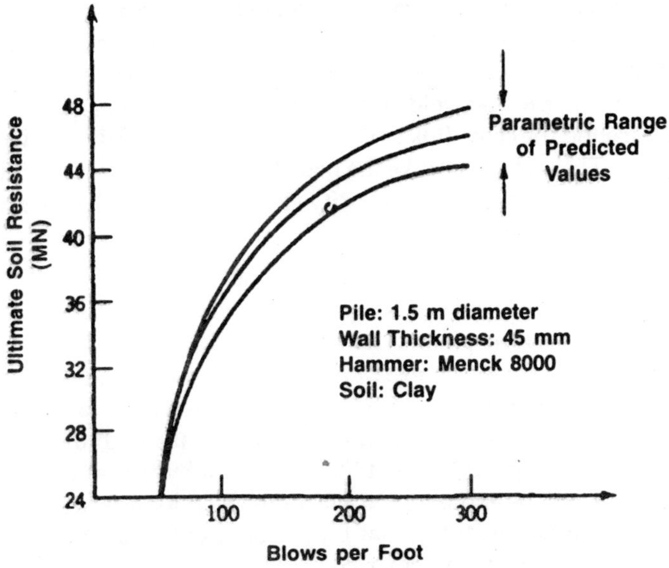


**Figure 7.5: Typical Wave Equation Analysis Prediction of Drivability**

solution has been to cap the lower end of the pile and to turn it in the water so that buoyancy provides support along the lower half of the pile. This maneuver requires consideration of water depth, pile length, net weight of the buoyant portion of the pile, and pile bending strength. Once turned to vertical, the bottom cap must be removed. The pile should first be filled with water to equalize the head on each side of the cap so that the closure plate may be safely removed. Other methods of removing caps have been used, in a variety of circumstances. On the Drift River Terminal in Cook Inlet, Alaska, lucite caps were used which shattered when the pile was driven down.

On the Hondo platform off Santa Barbara, California, a chain was welded in spiral fashion to a steel plate; the idea was that when the chain was pulled, it would rip open the plate like the cover of a sardine can. It worked in tests on land; it did not fully work in the actual installation, with the result that some plates had to be drilled out–a costly and time-consuming operation.

A similar failure occurred on an offshore terminal at Inchon, Korea. The steel plate closures were seal-welded and then a chain hinge provided at one side. The concept was that when the pile was driven down, the seal welds would break, the plate break free, and hinge around the chain link to move clear of the pile. It did not work that way. The steel pile broke the plate loose, but it then rode down as a plate cover to the pile, making it a closed end pile and preventing full penetration until drilled out.

Similar closures are used to keep the jacket legs buoyant during launching, floating, and upending. They are designed to be ruptured by the impact of the pile. Nylon-corded and reinforced neoprene closures are now used almost universally. They are domed to resist hydrostatic pressure efficiently, yet they are designed to be easily cut out like cookies by the pile as it is driven through them. The closures are designed to resist the maximum hydrostatic head to which they will be exposed. On a few rare occasions these closures have been made so strong that they could not be cut through by driving. They then had to be drilled out.

Rock drills are not efficient in drilling rubber; as one can imagine, the teeth become fouled and the water jets cannot clear the rubber. It requires many trips of the drill stem in and out of the pile in order to finally clear the seal. One solution, then, has been to cut the end of the pile on a scallop, like teeth, so that it slices through the seal progressively.

Reinforced rubber seals remain the state of the art despite the potential problems they can pose–for example, when drilling must be carried out through the pile and cut slabs of reinforced rubber are encountered, causing the drill to be clogged.

Diaphragm-type rubber leg closures are available in sizes from 18 to 144-in. O.D. For deep-water structures and very-large-diameter legs of jackets, sleeves, or cylinder piles, mechanically locked rubber diaphragm elements are available for pressures up to 14 MPa on 2.2 m O.D. closures and 2 MPa on 3.75 m O.D. sections. Special closure cutting tools are available.

Increasingly, the piles for deep-water offshore platforms are arranged in clusters around the corner jacket legs and their loading transferred to the jacket by means of sleeves bracketed out from the sides. The final top of these piles will then be underwater a distance equal to the water depth less the sleeve length. This latter is usually 20 to 30 m. The pile connection is made by grout.

To drive the pile so far below water requires the use of either an underwater hammer or a follower. Several types of hydraulic underwater hammers are now made, two of which can fit inside the pile guides which are bracketed out from the jacket at higher levels. Most common, however, is the use of a follower. This is a thick-walled pile section with a machined driving head on its tip which fits snugly over the head of the pile, transmitting axial compression, while preventing local buckling. For long followers, it may be necessary to join segments with mechanical joints, such as the breech-block connectors previously described.

Occasionally, due to misalignment or minor variances in the pile head, the pile becomes jammed into the driving head and the follower cannot be removed. Then the pile must be cut off, either by divers or else by a drill using expanding casing-cutter tools. To prevent excessive delays under such a circumstance, the corrective tools should be on board.

Experience shows that with a properly fitting driving head, a square cut on the pile, and a pile wall thickness that is not too small, that is, not less 25 mm, then there is very little loss in efficiency by use of the follower.

Where excessively hard driving is expected–as, for example, when driving though limestone or caprock strata–a driving shoe should be provided at the pile tip. API RP2A suggests that this be at least one diameter in length and have a wall thickness 1.5 times the minimum thickness of the parent pile section. Experience in driving through weak limestone containing embedded basalt cobbles has indicated that such a shoe should be two diameters in length to prevent buckling like an accordion. Steel quality should be as high yield as can be properly welded; since the weld is made in the shop, it can be properly preheated and post-treated as required.

## The Inflatable Packer and Grounting Arrangement

The packers must be installed at the bottom of the sleeve as to protect them during pile entry and driving.

Some designs of packers are passive, that is, just flexible rubber. Others are expanded by water pressure or by the grout itself. In the

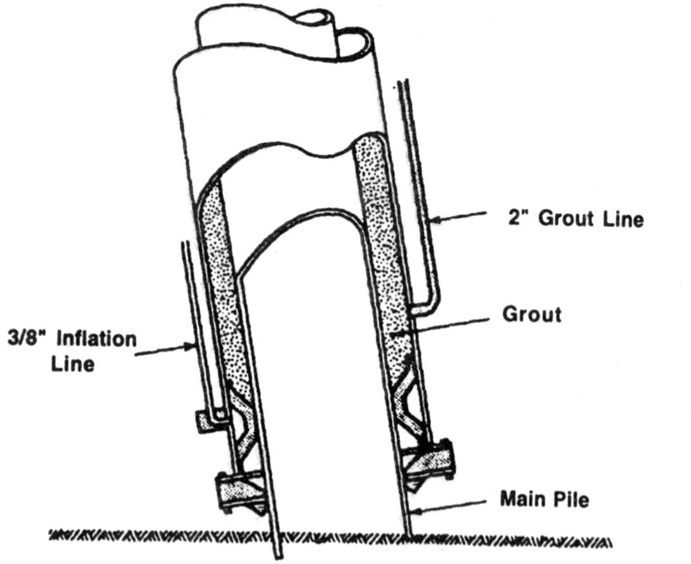


**Figure 7.6: Typical Inflatable Packer and Grouting Arrangement**

latter case, the grout first fills the packer; then as the back pressure rises, the grout opens a flap valve into the annulus. When a packer is damaged and the grout is escaping, then all that can be done is to allow the grout to set and try again. Unfortunately, it will tend to set in the grout pipe also. Flushing out slowly with water at minimum pressure can be used to keep the grout pipe open for the second injection.

The grout equipment should maintain continuous flow until the annulus is completely filled. If the configuration and relative elevations do not permit grout to be returned to the surface to verify complete filling, then suitable means should be employed, such as electric resistivity gauges, radioactive tracers, pneumo-fathometers devices, or overflow pipes which can be verified by divers or ROVs.

Recently, a new method of locking piles to sleeves has been developed in which the pile is "forged" into recesses in the sleeve by means of intense hydraulic pressure. This method, known as "HydraLok," has been successfully employed to fix pin piles to the sleeves of subsea templates of the Balmoral and Southeast Forties fields in the North Sea. It has recently been used at depths up to 1000 m. While it forms a reliable clamping, it is usually supplemented by grout for the permanent connection.

Pile installation records must be kept to record the following data:

1. Pile identification.
2. Lengths of each segment of pile.
3. Penetration of pile under its own weight, after penetrating the pile closure.
4. Penetration of pile under weight of hammer.
5. Blow counts throughout driving.
6. Unusual behavior of hammer or pile during driving; for example, a sudden decrease in resistance which is not explicable by a review of the soil profile may indicate a ruptured weld.
7. Interruptions during driving; a record of "setup" time and a record of the blows subsequently required to break the pile loose.

The Hondo platform was built in a water depth of 264 m. The soils are primarily finely grained, normally consolidated cohesive silts. There are eight main piles, up to 382 m in length; 12 skirt piles were driven. The main piles are 48 in. O.D., the skirt piles 54 in. O.D. Breech-block connectors were used for the followers for the skirt piles. The main piles required 13 add-on sections due to their long length. Hammer size was limited by the then-current availability to the Vulcan 3100 hammer, with a Menck 4600 hammer as a backup, both developing about 400 kN-m of energy per blow. Derrick boiler capacity was also a constraint. Extensive investigations and analyses were made to predict pile-driving performance. Based on prior experience, hammer efficiencies were assumed at between 55 and 80 per cent, with 80 per cent as the most probable. Pile-driving logs, hammer blow counts vs. driving resistance, and pile head forces were all generated for each assumed efficiency rating. Special attention was directed to the last three add-ons.

The *Thistle* platform in the United Kingdom sector of the North Sea was built in 161 m of water depth. Piles were designed for ultimate axial loads of 35 MN (3500 metric tons). These required penetrations up to 140 m below the seafloor into hard clay with multiple sand lenses. Wave equation analyses showed that only about 30-m penetration could be achieved by the available hammers. Therefore, a two-stage pile solution was selected. In the first stage, 1.37-m piles were driven to 30-m penetration using a Vulcan 560 hammer (400 kN-m energy per blow). The piles typically formed a plug at about 25-m penetration, when the driving resistance would rise from 150 to 250 to 600 blows/m. Some of the plugs could not be broken free by driving and had to be drilled out to enable further penetration. Therefore, a 1.5-m-long driving shoe of steel pipe 12 mm thicker than the normal wall was attached to the tip; this was beveled to force the soil out from the pile tip rather than inside. Piles were driven with a follower, assembled by breech-block connectors. A hydraulic clamping-aligning frame was used to hold the pile add-on. From 30 to 140 m penetration, holes 1.21 m in diameter were drilled. Salt water was used as the drilling fluid (with return to deck level at +30 m) until an 85-m penetration was achieved. Then because of interbedded sand lenses, drilling mud was employed. To prevent hydraulic fracturing in the sand lenses, the mud specific gravity

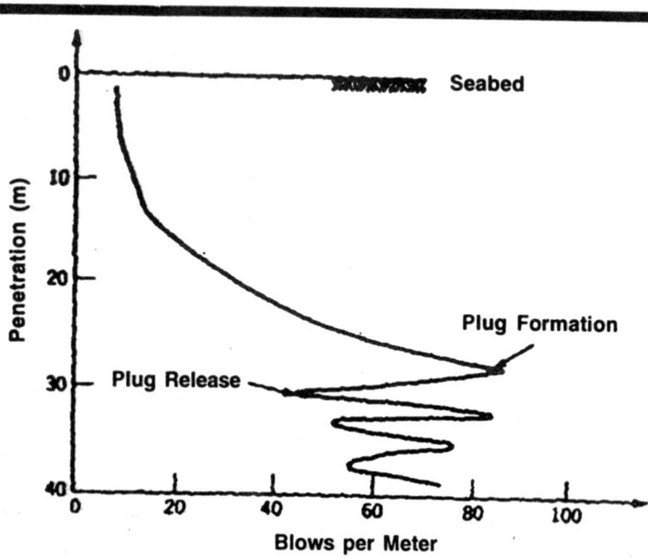


Figure 7.7: Typical Pile Penetration vs. Beiow Count in Hard Clay

had to be very carefully monitored and controlled. Holes were drilled with a flat bit, using airlift reverse circulation in a dual-walled drill pipe. When the holes had been drilled, they were gauged and found to vary from 1.22 m in diameter in clay to 1.52 m in sand lenses.

For the *Heather* platform, also in the United Kingdom sector of the North Sea, large piles were driven to extremely high capacities in hard, sandy, silty clay. Pile design loads were 29.5 MN, requiring an ultimate capacity of 44.3 MN. Pile penetrations of 43 m were required. Total pile length was 96 m. Piles were of 1.52 m diameter and used 64-mm walls throughout to enhance driveability.

Connection to the sleeves was by grout. To improve bond transfer, rings of weld beads were run on the piles. The first pile section was 64-m long, the second 32 m. To expedite add-on, a hydraulic clamping and aligning device was stabbed onto each first section. An internal driving shoe, 87-mm thick and 0.5 m in length, was attached, having the same outer diameter as the pile. Pile-driving performance was predicted by means of geotechnical investigation and use of the wave equation. Special attention was paid to the mechanism of plug formation. It was predicted that plugs, once formed, would give only partial end bearing under the hammer blows because of the different transmission velocities of the compressive wave. This was partially confirmed by the actual performance, in which blow counts built up to 500 blows/m and then stayed about constant for further penetration. A pile follower was used, made up of two sections. Gravity (machined for direct bearing) connectors were used between the follower sections and between the follower and pile. To achieve the very high capacities, Menck 8000 and Menck 12500 hammers were employed, the latter developing 2000 kN-m of energy per blow. Initial driving, however, was started with the smaller Vulcan 560 and Menck (400 n-m per blow) to reduce pile stresses. Pile performance was carefully monitored by means of strain transducers and accelerometers in the pile head. These showed that the losses in gravity connectors were only 2 per cent, except in one case of misalignment. Hammer efficiencies ranged from 40 to 62 per cent.

At the site of the *Maui* A platform, clayey soils were interbedded with a dense layer of volcanic ash. Some piles were able to penetrate this layer; others hung up at refusal. The problem was aggravated by hammer breakdowns and weather delays, leading to high setup. Eventually jetting was employed to break up and remove the plug of ash which had transformed the piles into end-bearing piles. A similar problem with volcanic ash occurred on a land piling project, where long steel pipe piles were being driven. Here also it became necessary to remove the ash plug, in this case by drilling out the pile plug.

Like many sub-Arctic and Arctic areas, the recent sediments of Cook Inlet are underlain by glacial till and overconsolidated glacially derived silts. The glacial till is similar to the boulder clay of the North Sea in being very stiff and containing rock fragments of various sizes. The overconsolidated silt is extremely difficult to penetrate under hammer impact alone; the material is so dense and its structure so strong that it can neither be displaced nor consolidated. Large cylinder piles from 2 to 4 m in diameter have been installed in the several Cook Inlet terminals by taking advantage of the rapid breakdown of the dense overconsolidated silt which occurs when water is introduced by jetting. A ring of jets has been built into the pile tip, arranged so that the jets can be continuously operated while the hammer works. This jetting then breaks up the overconsolidated silt, into a colloidal suspension. Powerful free-jets can also be used to wash holes in the dense soils, which then allows further penetration of the pile. Similar benefit to pile installation has been reported from the East Dock at Prudhoe Bay, where jetting enabled ready penetration of 25-m-long steel sheet piles through overconsolidated silt and partially ice-bonded sands.

## Handling of Offshore Piles

The handling and positioning of offshore piles involves number of problems, such as length, lack of fixed reference points etc.

Therefore, some form of template becomes necessary. The typical offshore terminal employs steel tubular piles, 0.6 to 1.5 m in diameter, with 1.0 m diameter most common. Wall thicknesses are 20 to 40 mm. Lengths are of the order of 40 to 60 m. Use of a permanent template, frame, or jacket gives the opportunity to make all intersecting welds under optimum conditions. Offshore structures are subject to cyclic dynamic loading and, hence, cumulative fatigue. Joints which are field-welded are seldom successful, due to the vibration during welding, the difficulty in keeping the metal dry, and the sudden cooling of the weld with seawater splash. Such joints are also subjected to corrosion-accelerated fatigue.

A template can be set on the seafloor, on the installation of steel platforms, or it can be suspended from an offshore barge or jack-up, or it can be cantilevered out from the previously constructed portion of the structure. In a few cases, self-floating templates have been used. Once one or more vertical piles have been driven, the support of the template may be transferred to these piles.

When piles are installed on a batter through templates near the sea surface, they are, of course, in cantilever through the water column until they are finally lowered far enough to engage the seafloor and obtain support at their tip. This situation frequently occurs in the construction of offshore terminals where inclined batter piles are used to provide the lateral resistance for mooring and breasting dolphins. The deflections may be significant in deep water (25 m) and may result in significant residual stresses in the piles. Various means of minimizing the deflections have been employed, including temporary evacuation of the water from the pile to provide near-neutral buoyant. Neoprene or frangible caps such as lucite can be installed on the pile tip, or the top of the pile may be capped and compressed air used to exclude water from entry at the tip. Removable floats can be attached to the tip, but these are awkward to install and are likely to break free. The above systems were all used on the VLCC terminal at Ise Bay, near Nagoya, Japan, because of the deep water (25 to 30 m).

Another way in which undesirable residual stresses are built into offshore piles is by releveling a jacket after some pin piles have been allowed to run in. The bending stresses so imposed may be significant. They led to serious problems requiring remedial action on several breasting dolphins at a terminal in Fao, Iraq. As noted earlier, good practice calls for initial setting of the jacket as level as possible, driving a few (three or four) pin piles just far enough to permit releveling and temporary fixing. Then after other piles are driven, the first piles are lifted back up into the jacket, clear of the seafloor, and then once again lowered and driven. Of course, if the initial setting of the jacket was level and needed no correction later, then the initial piles can be directly driven on down. Jacket-leveling devices are commercially available.

Large-diameter cylinder piles are often used for the mooring and breasting dolphins of offshore terminals. They are often delivered by self-flotation, with diaphragm closures. Ballasting may be used to help upend them. Because of their large size, the effect of waves and currents on them may be significant. In Cook Inlet with high-tidal-current velocities, vortex shedding caused oscillating transverse forces to act on the pile. This was in addition to the large direct force of the current. Guiding frames were required, cantilevered out from the derrick barge, with hydraulic positioning devices, in order to enable the piles to be set vertically in their correct position.

Installation of drilled and grouted piles may be carried out in either one or two stages. The piles are to be placed within drilled holes, which are held open temporarily either by seawater or drilling mud. A casing must first be placed through the water and seated into the soils to prevent flow under the tip and piping due to the imbalance in fluid heads during drilling. Usually this is accomplished by just driving the casing into the overlying soils to a moderate penetration. Then a hole is drilled ahead, using either direct or reverse circulation.

The progress of drilling depends primarily on the selection of the proper type of drill bit, suitable to the quality of the rock (hardness, toughness), and on the weight on the drill stem. The progress also depends on the ability to flush the cuttings form the teeth and to discharge them, *i.e.* adequate velocity and volume of the drilling fluid. If gravel is encountered, which cannot be flushed out, the particles will just roll around until they are finally ground to powder. Similar impediment to progress occurs if the fragments are not adequately flushed and discharged.

Since the piles and hence the casings for offshore piles are usually of relatively large diameter (*e.g.*, 1 to 2 m), normal direct circulation will not develop sufficient return velocity of flow to transport the, cuttings to the surface. The annulus between casing and drill stem is just too great in area. By building up the drill stem by means of pipe sections, the annulus can be reduced, thus increasing the return flow velocity to satisfactory velocities.

The relatively rapid flow of the fluids through the drill bits help to clean the teeth. Conversely, the flow may erode the walls of the hole below the tip of the casing. Hence, reverse circulation is most often employed. In this operation, seawater or drilling mud is pumped into the casing to keep its level at the desired head. The fluid then moves slowly down the casing and through the teeth and then accelerates to a high velocity up through the drill stem. This high velocity can remove cuttings of high density, such as pyrites. The low velocity along the walls of the hole prevents erosion and minimizes caving.

The head inside may be built up above the outside sea level; if carefully done to a pre-calculated differential head, this may help to keep the hole open. If the inside water level falls below the sea level, the hole may cave. One means of preventing this is to cut windows in the casing at the prescribed elevation above sea level. In most cases, seawater or slurry will have to be pumped into the pile, especially during airlifting.

When holes are drilled ahead of the casing, they are in effect cantilevered beyond the tip. Hence, they must have as much guidance as possible within the casing. Centralizers or stabilizers can be used on the drill string to keep it centered. A heavy weight on the drill bit will help to align the hole vertically. This can also be used below the tip of the casing in hard and firm soils and rock. When holes are drilled on a batter, there is a natural tendency for the hole to droop. This droop can be countered to some extent by the stabilizers and by reinforcement of the drill string to increase its bending stiffness.

Drilling muds can be used to keep the holes from caving and are especially useful in sandy soils. The drilling mud has a higher specific gravity and hence overbalances any inward flow of water. It also penetrates the sands and forms a cohesive layer. Using a bentonite slurry helps to keep the hole open.

To overcome this will require flushing of the hole prior to grouting. A thin drilling mud or seawater may be used for flushing. Alternatively, a polymer mud may be used throughout, which overcomes the majority of difficulties but is more expensive. It actually improves bond. The polymer slurries are non-toxic. With current environmental restrictions in most coastal areas of the world, it appears definitely advisable to consider the use of a polymer slurry.

It is essential to monitor carefully the fluid weight and head of drilling mud, since a few pounds per cubic foot can amount to a significant head differential at depth. Care must be taken in all drilling operations not to "fracture the formation" by excessive head of drilling fluid. In this case, piping ensues and the drilling fluid is lost to the sea. One method to prevent this is to "spot mud" only, that is, drill with saltwater at an equal or slightly higher hydrostatic head than ambient and then "spot" drilling mud in the socket only (not the full casing) to keep it open long enough to install the pile. If excess head is developed, resulting in formation fracture, it may become necessary to grout the hole with a weak grout, such as a sodium silicate foamed grout. After the grout has set, re-drilling in the same location must be done with care due to the tendency for the bit to wander off line as a result of differential cementation.

After the hole has been drilled, the pile is placed and grouted. It should be centered in the hole by spacers to preserve an annulus for grouting. As an alternative to inserting a pile, the drill string may be used as the pile by fitting expendable cutting tools to the tip to avoid the time required for removing the drill and insertion of the pile. This is especially effective for tension piles.

In the case of a two-stage drilled and grouted pile, the primary pile is driven to a predetermined tip elevation. This elevation is selected as one at which there is confidence that the pile can be driven and below which there is confidence that the hole can be kept open. This outer pile then becomes the casing for the drilling of the second-stage pile. When piles are placed in drilled holes (beyond the tip of the casing), the diameter of the drilled hole should be at least 150 mm larger than the outside pile diameter.

As described earlier, drilled and grouted piles using two stages were used very successfully on the Thistle platform in the North Sea. The primary piles were driven through the overlying sands and clays having interbedded sand lenses to seal in a dense clay stratum. This work could be carried out expeditiously during the short summer weather window. Then drill rigs were set on top to work during the winter, drilling ahead, placing secondary piles, and grouting the whole to act integrally with the jacket.

Drilled and grouted piles are especially effective in some calcareous sands. The grout penetrates the crushed shells and develops interlock with the unbroken shells behind. This appears to be the only positive method of achieving good skin friction transfer in calcareous sands. However, there are some calcareous sands, notably those on the Northwest Shelf of Australia, which are essentially impermeable and for which grouting just crushes the sand, increasing the effective pile diameter slightly but not increasing the effective friction significantly.

A grouting shoe may be installed near the tip of the pile in order to permit grouting of the annulus without filling the interior of the pile. A packer, of course, serves a similar function. With closed-end piles of large diameter, it is necessary to check that the pile will not be raised (floated) by the pressure of the grout as the fluid grout is placed. In soils which may soften or slack upon exposure to water, the pile should be placed and the grout injected as soon as practicable after drilling. The quality of the grout should be verified at frequent intervals during injection.

Holes for adjacent piles, for example adjacent skirt piles under one leg, should not be open at the same time unless the soil properties are sufficiently strong and consistent to ensure that grout will not migrate during placement into the adjoining hole through fractures or seams and that the soils will not suffer relaxation.

There are several types of drills available for use, depending on the size and depth of hole and the available supports for the drill. For emergency or limited use, such as arises when a single pile encounters a boulder, a churn drill may be used, suspended from the crane boom if necessary. Another type of drill, a pile-top rotary drill, is supported on the pile or casing itself, and gets its torque reaction from being clamped to the pile. Drilling subs and drilling swivels are extremely versatile and easily handled drilling machines. They are suspended from the derrick boom. They get their torque reaction from chains attached to an arm. This arm and the chains are subjected to very severe shock loads and hence must be properly designed and secured for impact. This type of drill is very flexible in use, since from one location of the derrick, several piles can be drilled. Most recently, down-the-hole pneumatic hammers have become available and are proving both effective and flexible in their use. They are particularly useful in fractured rock whereas rotary drills, may have difficulty in setting started. The Calyx system of drilling uses tricone bits mounted on the periphery of the caisson tip. These drills can drill a vertical hole to 1/2 accuracy. The drills are steerable.

For major pile installations the use of a drilling rig with inclining mast appears to be most applicable, since it can attain faster drilling rates and make and unmake long drill strings more rapidly. Such a drill rig is usually skid-mounted and is set on a temporary work deck mounted on the jacket. These drills use conventional rotary drill bits. They may be operated either in direct circulation or reverse circulation although the latter is generally most satisfactory.

## Bell Footing for Platforms

One important development in pin pile installation of offshore platforms has been the bell footing, first used on the Ekofisk Field platforms and since extended to other structures in the Arabian Gulf, the Northwest Shelf of Australia, and Southern California. In this case, the primary pile serves as a casing through which a drill rig drills a moderate-length hole ahead. Then it employs a belling tool to enlarge the socket into a bell of 4- to 5-m diameter. Reverse circulation is employed, usually with a bentonite slurry (drilling mud) as the drilling fluid. Then a heavy reinforcing cage or steel insert pile is set. The bell and socket and a portion of the casing are filled with underwater concrete, using "fine concrete" aggregates, for example, maximum size 9 mm. As with straight drilled shafts

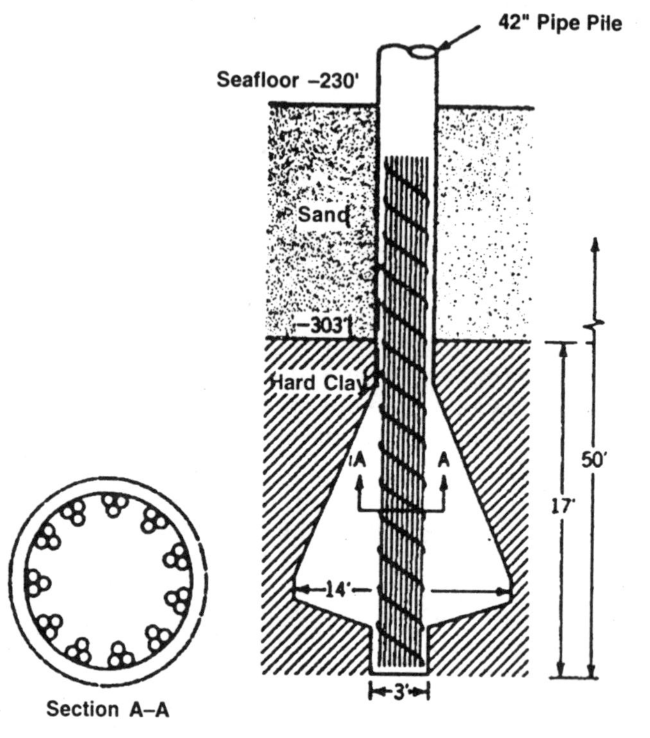


Figure 7.8: Underreamed Bell footing for Ekofisk Platforms

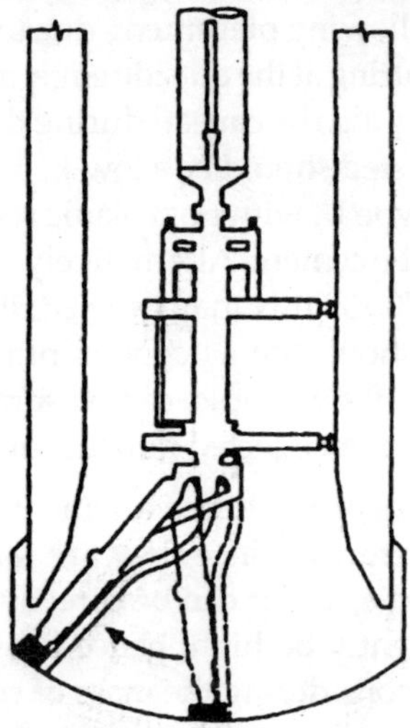

**Figure 7.9: Mitsui "Aqua-Header" Special Drilling System for Large-diameter Cylinder Piles**

(sockets), saltwater may be used as the drilling fluid and the bell "spot-mudded" with polymer mud to hold it open until concreting.

API Standard RP2A describes belled piles as they are used to give increased bearing and uplift capacity through direct bearing on the soil. A pilot hole is first drilled below the base of the driven pile to the elevation of the bell base and slightly below to act as a sump for unrecoverable cuttings. Then the bell is drilled, using an expander tool. Reverse circulation must be employed in order to gain enough discharge velocity to remove the cuttings. Slurry is usually used to keep the bell from collapsing; alternatively, the sands surrounding the bell can be stabilized by epoxy injection.

It is not possible to confine the bell concrete with hoop reinforcing. There is no way the hoop steel can be placed. If the bell has been drilled into rock, the rock may confine it; however, bells at both Ekofisk and North Rankin A platforms were used in soils of relatively low stiffness compared to the concrete. Hence, there will be flexural and shear stresses in the concrete under service and extreme loading conditions. This means that the shear and tensile strength of the concrete must be utilized.

Such a confined mass of concrete will get very hot during the period immediately following placement, due to heat of hydration. Subsequent cooling, starting at the outside edge, may produce severe cracking. Cracking may also be caused during the expansion stage. Hence, the cement selected should be a low-heat-type cement, such as ASTM Type IV or Type II, with pozzolanic replacement of up to 30 per cent or more of the cement. Alternatively, coarse ground slag-portland cement in a 70 : 30 mix may be used; it has very low heat generation. The mix should be as cool as practicable at time of placement; aggregates, for example, can be sprayed with water to cool them or liquid nitrogen injected into the mix.

Drilling contractors prefer to use a cement slurry (*i.e.*, cement plus water plus admixtures), since they are familiar with it, it is relative simple to handle, and it can be directly pumped. Cement contents will consequently be high; hence heat of hydration is a very serious problem considering the mass of concrete in a drilled shaft. Temperatures exceeding boiling and disruption of the concrete have occurred. Use of blast furnace slag-cement is indicated. However, the tensile strength and shear strengths of neat cement slurry (grout) are relatively low. Steel fibers have been proposed as one means of enhancing the tensile strength. However, they may tend to segregate in a cement slurry. The use of polypropylene fibers will give more satisfactory behavior. A mix which incorporates sand-that is, a sand-cement mortar- or, even better, one incorporating small aggregates (8 to 10 mm) such as pea gravel will have relatively low heat properties and good tensile and shear strengths, yet will be able to flow readily. Plasticizing admixtures should be used. If silica fume or antiwashout admixtures are added, then a superplasticizer high-range water reducer is also required as well as a retarding admixture to prevent the superplasticizer from a sudden loss of slump due to the heat generated.

Use of valves or restrictions at the tip of the tremie should be avoided. If necessary due to other aspects of the operations, the valve must be carefully designed so as not to jam with a piece of aggregate, and must not abrade under the flow of concrete. Several devices have been developed for this purpose; the most successful involve hydraulically operated squeeze valves.

As described previously, to start the placement, the tip of the tremie pipe is closed with a gasketed plate, wired onto the end of the pipe. The pipe is assembled and run down to the tip. Then the pipe is charged, initially with the 1 $m^3$ of cement grout, then filled to slightly above half-way. The pipe is now raised a few inches above the bottom, allowing the plate seal to break free and the concrete to flow out as slowly as it can be controlled, until it reaches a balance of heads. Then concrete is continuously fed in at the top. At greater depths, a polyurethane pig may be employed, being pushed down the casing by the initial flow of concrete. Using a pig, tremie concrete was successfully placed in belled footings under the primary piles of the North Rankin A platform, at a depth of 240 m below sea level.

The annulus between insert pile (or reinforcing cage) and primary, pile must be 6 to 10 times the maximum size of coarse aggregate to permit flow up around the insert pile.

Electrical resistivity or well-logging devices can be used to determine the level of the top surface as it rises up in the primary pile. Obviously, the placement should be extended several meters above the design level to account for any mixing of the initially placed grout with the drilling mud. Bentonite drilling mud should preferably be converted to a calcium base to a avoid coagulation upon contact by cement or even better, a polymer mud used.

Although drilling contractors and others continue to urge placement by pumping, their experience is primarily been with the cementing of oil field casing and with grouting of annuli. For the much-larger-diameter drilled shafts and belled footings, experience conclusively shows that reliable placement can only be attained by gravity feed.

The belled footings of the North Rankin A were constructed in a weak calcarenite. An elaborate stabilization process was used, by which the surrounding sandy matrix was impregnated with a thin epoxy. Then the bells were drilled, an insert pile placed, and the bell and pile filled with tremie concrete placed by gravity. Because the surrounding soils and seawater were at an ambient temperature of 38°C, the mix was precooled by liquid nitrogen to 5°C. Cold water was flushed down the 90-mm dia. tremie tube before the concreting started. Coarse aggregate was 8 mm maximum. Subsequent coring showed concrete tensile strengths of 6 to 7 MPa and compressive strengths about 60 MPa.

Once piles are installed, it is necessary to evaluate their load-bearing capacity to ensure that the required capacity has been attained. Sometimes the skin friction may be deficient; this is frequently encountered in calcareous soils. Another situation that may arise is where an existing platform is to be upgraded to withstand greater environmental or operating loads.

One method is to clean out the internal soils to a safe distance above the tip (usually several meters) and then to construct a concrete plug inside the pile tip, of adequate length to develop bond transfer to the pile. This converts the pile to an end-bearing pile. This method was successfully employed for the previously driven piles in the Kingfish platforms A and B in Bass Straits, Australia, which were driven into calcareous sands. On the North Rankin A flarestack, also in calcareous sands, after the concrete plug was placed, grout was injected under the plug to reconsolidate the sands.

When inadequate capacity is encountered in the initial driving and the pile fails to develop adequate resistance at the design penetration, then appropriate steps taken promptly may enable the construction of an adequate foundation with minimal additional cost. For example, on one offshore terminal in the Mediterranean, the 2-m-diameter open-ended steel piles failed to develop the required resistance with 60-m penetration. On the first pile it was found that the penetration would have to be increased by more than double, at a prohibitive cost. By welding on a tip closure plate which closed 80 per cent of the tip, leaving a small central hole for water escape and relief of soil resistance, the piles developed adequate static capacity in the calcereous sands and safely sustained a test load of 4000 tons.

When increased tension capacity is required, two methods are possible. One is to drill in an insert pile. In one case the drill string itself was used as the insert pile and grouted into place. Another solution is to weight the pile, similar to placing weights on the legs of a table. The pile is cleaned out, a concrete plug is placed, and iron sand or barites are placed. These should be carefully selected for in-place density, durability, and freedom from corroding effects on the steel. In-place densities of 3.5 have been achieved with iron sands. Both systems were employed for the piles of the Kingfish A and B platforms in Bass Strait, Australia.

A third solution, applicable in stratified soils where both compression and tension capacities must be increased, is to construct a belled footing, as described above.

Insert piles may be driven through existing piling, being made up in short segments, welded as they are installed. They are driven with a follower using mechanical threaded connections for the follower segments. After driving below the original tip a sufficient distance to develop the required bearing, the insert piles are connected to the primary piles by grouting of the annulus. Alternatively, the insert piles may be installed by drilling and grouting.

Lateral resistance of existing piles may be increased by installation of an insert pile, grouted or concreted in, to stiffen the pile in the vicinity of the seafloor. This was successfully carried out for a platform in The Bombay High field, India, where the pile was showing excessive deflection under cyclic loads. The soils were calcaeous sands. The seafloor sediments surrounding the pile may be in some cases strengthened by vibratory densification or pressure grouting, or they may be surcharged with gravel, either normal density or high density, to consolidate the existing soil, replace any settlements, and fill any gaps that occur under cyclic wave action. Stiffness of the pile may also be increased up to 25 per cent or even 33 per cent by filling with concrete.

It has long been the practice with tubular piles for bridge piers to clean them out and fill them with concrete, in order to improve both axial and lateral capacity. Reinforcing steel cages are usually installed. To prevent loosening of the soil around the tip, during clean-out, the pile should be kept full of water and the removal should stop one or two diameters above the tip. To prevent settlement under high loads, the pile tip should be grouted under pressure after the concrete fill has set. Composite action between steel and concrete core can be obtained by shear rings on the inside of the pile head.

Freezing of the soil around the pile and under the pile tip is another method which has been proposed as a means of increasing the capacity of previously driven piles. The concept is that, after driving, the steel pile would be cleaned out by jetting, airlifting, or drilling and then used as the freeze pipe casing.

## Chapter 8

# Breakwater Structures and Offshore Terminals

## Introduction

The breakwater structures vary depending on the wave climate, the slope of the beach and the depth of the breakwater. The designs of breakwaters are becoming increasingly sophisticated as a result of advanced understanding of hydrodynamics of wave interaction with the structure and the sloping bottom. Experience has been accumulated worldwide and translated into these improved but more complex designs. For the constructor, this means the positioning and placement are very demanding, especially in an environment where the waves are typically steepening and refracting.

Concurrently, there have been significant improvements in methods of control and survey. Crane barges are larger, with more stability. Cranes have greater reach. Mooring systems are available with increased holding capacity for taut-line moors. Buckets (grabs) can now be fitted with electronic position indicators connected to GPS or DGPS and read out in the operator's cab. Acoustical profilers can give real-time two-dimensional cross sections, automatically corrected for roll, pitch, and tidal variations, while photogrammetry can give accurate three-dimensional pictures of the above-water portions.

Now let us try to understand some information about offshore terminals. Offshore terminals are typically built in water depths exceeding 20 m in order to accommodate very large crude carriers (VLCCs) and deep-draft ore carriers. Wherever feasible, of course, these have been located in protected or semiprotected waters, but on many continental margins adequate water depth is found only offshore, in a partially or fully exposed location. Therefore, the construction operations must be carried out in the ocean environment, subject to the normal waves, wind, and current for that season, and with suitable precautions for possible storms.

## Types of Breakwaters

### Rubble-Mound Breakwaters

These are usually constructed with a core of quarry-run rock, overlain by one or two layers of larger rock carefully sized to prevent leaching out of the fines from the core. On the top and sea sides, large riprap or concrete armor units are installed.

In selecting the quarry for production of the required rock, primary consideration has to be given to ensuring that the requirements as to durability and abrasion resistance can be met. Rock will be required in different sizes: the most critical will be the large armor rock. The development and blasting plans for the quarry have to ensure that an adequate quantity of each size can be obtained. These plans must include temporary roads so that the hauling to the sorting and stockpiling areas can be carried out efficiently.

Blasting will usually be done with widely spaced, large-diameter holes, and slow powder, such as ammonium nitrate, in order to produce large individual pieces for the armor stone. The other rock is put through a grizzly (grid) in order to separate the Class B rock, which is the next largest size. Subsequent grizzlies used during loading can remove unwanted fines from the core rock.

Rock tends to break along pre-existing fractures. Examination of the quarry face will indicate whether the stone will break in properly rounded shape, or in slabs. Abrasion is normally determined by the Los Angeles Rattler Test. The handling of rock in the quarry is normally by loader, although a crane may be used to separate and stockpile the armor rock. Care must be taken in handling and loading the larger rock to prevent excessive breakage of the edges and corners. Transport of the rock is either by barge or truck. When a barge is used, the deck is protected by timbers.

Where the core is placed on a sand seabed, either small stone particles must first be laid or a filter fabric, to prevent the sands from migrating up into the interstices of the material above. During storms, excess pore pressures develop in the underlying sands, causing local liquefaction with the consequence that if a proper filter has not been placed, the breakwater "sinks" into the sand.

Filter fabric is a good alternative to stone but requires special means in order to place it and to keep it temporarily in place. It can be fixed to large frames made of steel angles, which stay in place, or made up as a mattress, with articulated concrete blocks attached. This latter is the practice in The Netherlands, where they first roll up the mattress on large buoyant steel drums, then later unroll it on the seafloor.

The core may be placed by bottom-dump or side-dump barge. Tests and experience show that lateral dispersion and segregation may be greatly reduced by pre-saturating the material on the barge to eliminate the entrapped air. Placement in a mass also tends to prevent segregation, although the impact in deep water will cause lateral windrows to form. The core can also be pushed off barges but the smaller rocks tend to glide laterally and segregate, whereas the larger stones go straight down.

Designs often call for relatively thin layers of progressively larger stones. In some wave climates, accurate control of these separate layers of different gradations becomes almost impossible. Pre-blending of two different layers may then be done, with the approval of the engineer. The layers then tend to automatically grade from finer inside to coarser outside provided the blended gradation is placed through a pipe or lowered to the seafloor in a bucket or skip. The next layer, usually referred to as Class B rock, typically 400 mm maximum, may best be placed by a skip or net although a large bucket is also used. Finally, the largest rock is placed, typically by bucket from a crane.

Because the maximum obtainable size of riprap is limited, breakwaters on exposed coasts often use precast concrete armor units. These armor units are designed to be stable against roll and displacement by the waves, and at the same time, to give a porosity that will dissipate the wave energy. A great deal of experimental research in both laboratory and field trials has been carried out. As a result, a Corps of Engineers' Manual lists over 40 different configurations. In recent years, much attention has been directed toward the actual performance under severe storms. It has been found that many of these sophisticated shapes tend to break up into large fragments, which then act to batter the remaining units.

It was found that much of the cracking originated during the casting, due to thermal strains and the restraint of the forms. As a result, the steel forms have been redesigned to accommodate this shrinkage during subsequent curing but the friction of the support still causes significant strains. Many tests and actual installations have been made with the use of reinforcing steel. This has served to pull the cracks closed and to hold the legs of the unit to the body, but has not completely solved the problem. Other steps which should help would be the use of blast furnace slag cement and/or fly ash, to reduce the heat of hydration, pre-cooling of the concrete mix, and insulation of the steel forms and the unit itself during the curing period. Due to the thickness of the members, drying shrinkage is a minor problem which can be adequately controlled by membrane curing compound.

During concreting of shaped units containing reinforcing steel, there is a tendency of the steel to displace upward, *i.e.*, to "float" during placement and vibration. This results in the steel being either exposed or having insufficient cover to prevent corrosion. Concrete dobe blocks should be fitted to hold the bars in place, both from their own deadweight and from uplift. Steel fibers have also been used to improve the apparent tensile strength of the concrete with relatively good results. However, the quantity of fibers required makes the costs very high. As a result of these problems, Dutch engineers have returned to the use of solid rectangular prisms, each weighing 1 MN (100 tons). These are randomly placed on the slope.

Concrete armor units are placed with slings, so that the unit may be oriented and placed with precision working from the lower end upward. They are usually placed with one or two segments per layer. The slings are fitted with releases that can be tripped from above. A single line sling is used to a drilled-in rock anchor, with a bridle so that the positioning may be visually controlled from above the water. When laying from floating equipment, a large crane barge is positioned inside the working end to be partially protected from the seas and to minimize the risk of grounding.

Survey control of the rock mound breakwater presents certain difficulty due to the fact that the critical points are under a zone of breaking and reflecting waves. Transverse profiling can be run on major breakwaters by side-scan sonar and a subsea profiler. Aerial photography can give contours of the above water sections. On smaller breakwaters, more conventional means can be used. A crane on top can reach out to hold a lead line with a heavy sounding lead, large enough so as not to go down a crevice. The line is located in the horizontal plane by conventional theodolite. Further off the centreline, helicopters can be used in a similar fashion. The lead line should have a weak link so that if it gets caught, it will not endanger the crane boom or the helicopter. If the sea is calm, then, of course, a boat can be used with both lead line and sonar. Sonar must be narrow beam and, even then, correction must be estimated for the slope of the beam since the sonar will measure the [illegible]earest point in its cone.

For breakwaters which must traverse sha[illegible] water or tidal zones where floating equipment cannot work, [illegible] access is "over-the-top" employing a trestle or working on top o[illegible] wi[illegible]ned section of core. Provision of sufficient width for the crane on top often requires additional material in the breakwater, not only core rock but also in the other layers over the cap as well. Core rock and stones are transported over the core by large truck. At intermittent locations, the core is widened still further to provide a turnaround or a turntable is provided, so that the truck can back out underneath the crane.

When trestles are employed, the crane may be mounted on a rail undercarriage and the rock delivered by rail.

Frequently, concrete caps are constructed on top of the breakwater. These must have adequate relief holes since otherwise the hydraulic ram effect of plunging waves on the open interstices below will generate large internal pressures that will blow the cap off, as it did on the initial installation of the breakwater for cooling water intake of the Diablo Canyon nuclear power plant in Southern California.

Similarly, the interstices in the above-water outer layers are often chinked with smaller stones; grout or concrete is then placed to prevent rats. Sufficient pressure-relief holes must be left open or otherwise they will open themselves during a storm with disruption of the surrounding stone. For placement of grout or concrete in the tidal zone, silica fume admixture gives both stiffening and antisegregation capabilities, that is, the concrete is more cohesive. Antiwashout admixture will prevent segregation but the lengthened set time may lead to washout.

The outer corners and exposed bends in the breakwater receive much more damaging attack, in part by wave concentration but also aggravated by the wave-induced currents. Additional armor and a thicker cross section are usually adopted.

The stability of individual rocks in the breakwater is a function of their specific gravity and is proportional to the cube of the underwater (buoyant) weight. Thus trap rock, with a unit weight of 3100 kg/m$^3$ (195 lb/cu. ft.) is more than twice as stable as normal siliceous rock at a unit weight of 2600 kg/m$^3$ (165 lb/cu. ft.).

## Concrete Caissons

Concrete caissons are frequently installed on top of a prepared stone bed. Typical segments, rectangular in cross section and from 20 to 60 m in length, are manufactured in a fabrication facility, launched or lifted out, transported and installed in much the same manner as previously described for quay walls, bridge piers, and submerged tubes. Care must be taken in the installation to ensure that they are temporarily supported at designed locations. One concrete caisson for the Tarsiut caisson-retained island in the Beaufort Sea was inadvertently seated with a large overhang, due to a surveying error. This resulted in negative moment and shear cracks, which fortunately did not prevent its successful functioning.

The major new element is the sealing of the gaps between adjoining caissons. The caisson-to-caisson tolerances have six degrees of freedom, so seals or gates must be designed to accommodate such differentials in each dimension and attitude. The best method is to install sheet pile arcs across each gap, one on the inside face, one on the outside face, and to fill between with stones, using a filter course on the bottom if needed to prevent washout in the inherent large turbulence that will occur at locations.

A sheet pile can be half-embedded in each opposing corner, using the heaviest flat sheet piles with the strongest interlock. These are anchored back into the concrete by welded reinforcing bars. Then an arc, similar to that employed in sheet pile cellular construction, is installed. Such caissons are employed on very large and deep breakwaters to form the core, with additional core rock, Class B and A rock and concrete armor units installed on the outer face.

Installations off the coast of Brazil and in the Bering Sea have proved almost unbuildable by the conventional system of setting sheet piles, one at a time, around a temporary ring wale or wales. In the case of Brazil, the solution adopted was to first make up a complete cell on a barge, with several ring wales joined in a space frame. The sheet piles were temporarily fixed to the frame and wales. This entire cell was then set by a large offshore crane barge and immediately filled part way with sand. Then the piles were released, a few at a time, and driven down into the underlying sands. The Bering Sea case was more easily solved by making the frame and wales much stronger and, extending it to give support to the sheet piles over the full length. The sheet piles were tied against inward and outward movement. A similar problem may arise in strong currents where an incomplete cell may be subject to the full current force, prior to the cell obtaining stability as a gravity structure.

Prestressed concrete sheet piles have been used to construct a number of breakwaters within harbors, in order to give protection to small craft such as fishing boats and pleasure boats. These sheet piles have to develop considerable bending moment, hence are usually thick, wide, and heavy as compared to conventional foundation piles. They are often prestressed eccentrically, which means that they may not have sufficient resistance against tensile rebound cracking during driving.

These piles have to have sturdy support during installation. They have to be set vertically and tightly against one another. To accomplish this latter, the tip of the sheet pile should be angled so that, as it is driven, it forces the toe back against the preceding pile. At the same time, the top, which is above water, is pulled back against the preceding pile. One method is to have a roller with two lines and an hydraulic "Turfur" hoist.

Sealing of the joints between sheets is often required to prevent sand from leaching out due to tidal changes and waves. The groove formed between the two sheet piles is cleaned out by jetting. Then a polyethylene tube is pushed down and filled with grout. Alternatively, a stiff grout with antiwashout admixture is forced down the grooves, to both seal and bond with the sheets.

## The Offshore Terminals

The typical offshore terminal consists of a loading platform, two (or four) large breasting dolphins, and four mooring dolphins. Catwalks usually join all these structures and may require intermediate supports. A trestle may connect the loading platform to shore, or submarine pipelines may be used instead.

The initial structural concept employed was an extension of the harbor-type structure of independent piles supporting a deck and fender system, adapting dimensions to the more severe design conditions of the open sea. Such structures have been extensively used in Japan, in the Arabian Gulf, and along the coasts of Brazil and Australia, for example.

Pilings are typically large-diameter steel pipe piles, 0.75 to 1 m in diameter, 40 to 60 m in length, having wall thicknesses of up to 50 mm. High-yield steel (350 MPa) is normally employed. Both vertical piles and batter piles are employed, intersecting at the deck level to react against each other under lateral loads. Batter piles are also called raker piles. As the size of ships and the exposure conditions become more severe, the proportion of batter piles

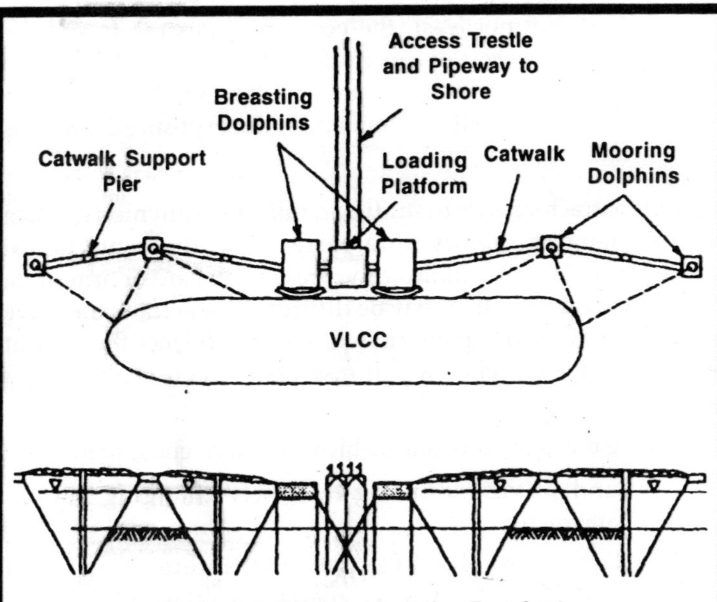


**Figure 8.1: Offshore Loading Terminal**

increased so that typically they dominates the construction. Relatively high axial loads are used as the basis for design, 400 to 600 tons in compression and 50 to 100 per cent of that in tension. Therefore, it is important that the piles be driven to their design penetration and that they be accurately located so that the axes of a batter-vertical pile group intersect at a single point, this latter to minimize bending. The connection at the intersection must be adequate to develop the flow of forces.

The construction problems for offshore terminals arise principally from the following:

1. The offshore site is usually moderately remote, involving logistical, personnel transfer, and survey problems.
2. The piles are large by harbor standards (although not if compared to those of deep-ocean production platforms).
3. Very close tolerances are required for positioning of the piles at their heads.
4. The relatively shallow water (20 to 30 m) causes refraction of waves and changes the wave and swell characteristics, making operations difficult even in moderate sea conditions.
5. Last, but not least, the economic constraints and the relatively short distance offshore tempt the contractor to utilize harbor-type equipment and methods, which often prove too small and limited for the exposure conditions and work involved.

The contractor needs to study carefully the bathymetry, which may be changing rapidly on a steep coast; the geotechnical information, which may indicate the presence of hard or firm strata or boulders and coral that may be difficult to penetrate; the wave refraction and breaking patterns, which may influence the work at the site; and the currents, which may affect the route of supply services.

Among the many problems which have been encountered are:

1. Sand waves, resulting in slowly changing bottom bathymetry.
2. Refraction of waves causing intersections and concentrations of wave energy at certain zones, with consequent pyramidal waves and confused seas.

3. Opposing currents causing steeper and higher waves.
4. Caprock on or near the surface that is difficult to break up and to penetrate yet is underlain by loose, almost liquid, sands and silts.
5. Weathered rock, which gives widely varying resistance to piles, even to adjacent piles.
6. Boulders on and under the surface.
7. Sloping hard surfaces, on or below the seafloor, causing the pile tips to tend to run downhill during driving.
8. Calcareous sands, requiring special methods in order to develop uplift and bearing capacity.
9. Overconsolidated slits and clays, which are extremely difficult to penetrate.

An initial construction requirement is to set up a shore base for support of the construction. Ideally, there will be an adjacent harbor with dock and craneage facilities. Unfortunately, this is usually not the case, therefore, such a support base must be established.

In the case of most offshore terminals, there are a very large number of structural and mechanical elements to be transported out to the construction site. An analysis of the number of barge loads and lifts will usually indicate that efficient transfer at the shore base is essential. Some protection must be provided against the breaking waves. Swells not only lead to much larger breaking waves, even in calm, winds, but also can lead to severe surges in harbors and channels, where mooring lines may be suddenly snapped. Adequate draft must be provided not only for the barges but also for the tug and crew boats. Services (power, water, and fuel) must be provided to the shore base, as well as communications, both local and to centers of supply, and especially to weather-forecasting services. Unfortunately, in the past history of offshore terminal construction, the contractor has often failed to set up an adequate shore base initially and has had to progressively improve it as the job went on, meanwhile suffering from the inadequacy.

The second step is to set up survey control. Horizontal control may consist of ranges, using focused brilliant lights visible for several miles at sea, even in the daytime, or lasers. Electronic-positioning systems are usually also installed. The shore stations are established on the coast. Tidal gauges are installed, preferably in protected wells. Levels can be run by laser, corrected for curvature of the Earth if the structure is distant offshore. GPS may be used as a verification of position. If real-time rapid satellite positioning is desired, differential GPS systems may be used.

As so often happens with all types of offshore structures, the geotechnical investigation made by design engineers for their purposes may be inadequate for the constructor's needs. Therefore, additional site information may need to be developed regarding the seafloor sediments, boulders, and obstructions. Through use of a "sparker" survey, jet probings, and borings, more information can be obtained on the upper soil strata through which the piles must be driven. For example, the contractor may want to handle and set a 60-m-long steel pile as a single piece on its designed batter. The contractor needs to have a reliable estimate of how far the pile will run down under its own weight. Will jetting be required? At the construction site proper, the contractor will now set up moorings, to facilitate moving the floating vessels and derrick barges along and around the terminal. Pre-set moorings minimize the time of moving and the problem of handling and resetting anchors. In many cases they eliminate the problem of crossed anchor lines when two pieces of floating equipment must work in close proximity.

Another early step may be the construction of an offshore survey tower, which will provide visual reference for close-in surveying. The availability of electronic-positioning devices of high resolution has minimized but not eliminated the need for such towers in recent years. Their ability to act as visual guides enables the crane barge superintendent to move rapidly to approximate position.

The next decision is what to do in case of storm. Presumably the boats will run to safety in a harbor. If reasonable weather-forecasting services exist, with proper planning and judgment the contractor may avoid being caught with supply barges at the site. But what about the major floating equipment, for example, the large derrick barge? This will always be at the site, and hence may be vulnerable to a sudden storm. It will not always be practicable or safe to try to tow it to a harbor; the harbor entrance may have breaking seas, shoals, or cross currents which make it exceedingly dangerous to enter in a storm. The tugboats may not have enough power and size to handle a large derrick barge under severe sea conditions.

Experience has shown that it is often safer to ride the storm out at sea on a pre-set storm mooring. Such a mooring will consist typically of a single long catenary wire line from the barge to a mooring buoy (spring buoy), which in turn has a line or chain leading to a heavy anchor. The anchor is usually two anchors, piggybacked (one behind the other joined by a halfshot of chain), or one large anchor, with a shot of chain to ensure that the pull is horizontal, "Moorings and Anchors." During any season when a sudden storm is possible, the storm mooring line is kept connected, but usually slack, to permit normal maneuvering on the short, taut operating lines. When the taut moorings are slacked, then the derrick barge will ride to the storm anchor. An auxiliary anchor, dropped underfoot off the bar, will prevent excessive yaw.

When construction proper begins, the piles are delivered, either on a barge or self-floating, and are picked up for driving. The first pile in each dolphin should preferably be one of the vertical piles in order to give a point of reference and support for subsequent batter piles. Yet many mooring dolphins are designed with all batter piles–no vertical piles!

A second problem is that the piles must be driven alongside each other and then cut and pulled to final position. With heavy-walled, large-diameter pipe piles which are driven, for example, through hard seafloor material, they may be quite inflexible and difficult to position. Pulling the heads may impose bending stresses that will be permanently fixed in the piles. A temporary vertical pile may be the most practicable solution for both problems.

There is another serious problem with the subsequent connecting of the piles. If these tubular members have to be cut, fitted, and welded at their point of intersection, this may prove impracticable to carry out in a satisfactory manner. The piles may be vibrating due to waves and vortex shedding from the current. The joint area may be wet with spray, the steel below optimum welding temperature. Weld positions will be unfavorable. These joints are subject to cyclic loading and dynamic loads; thus they often fail from fatigue.

As a result of the twin problems of positioning batter piles and connections, the prefabricated template has been evolved. In this case, a template, typically extending from high-tide level up to the top of the dolphin (a height of 5 m or more), is prefabricated of tubular steel members. All welding is done in the shop, where conditions are optimum and proper procedures and nondestructive testing can be carried out. The template will have sleeves, at the proper angle, through which the piles may be set and temporarily supported. The template is then held over the stern of the derrick barge in proper plan location, and one or more vertical piles driven through their sleeves. They are not normally driven to final penetration at this stage but only as needed to provide lateral and vertical support to the template. If the final design does not contain a vertical pile, then the contractor adds one, with its sleeve, as a temporary support pile, usually in the center of the dolphin. Now the template is raised vertically to proper elevation and temporarily welded off to the vertical piles. Its position and orientation are checked.

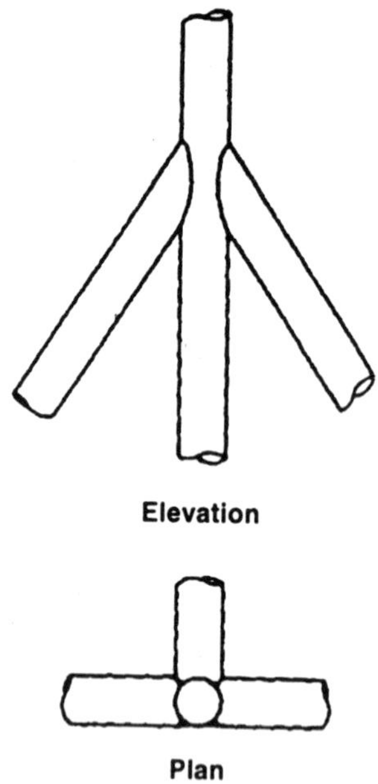


**Figure 8.2: Typical Intersection of Vertical and Batter Piles of Mooring Dolphins Indicating Difficulties Involved in Making Proper Connections**

Batter piles are successively set through the sleeves, alternating directions to avoid dislocating the template. As they are set, they are driven to grade. Finally, the vertical pile is cut loose, spliced, and driven to its final penetration. If the vertical pile is not in the design, but was only supplied by the contractor for temporary purposes, then it is cut loose and removed.

The connections between piles and sleeves of the template are usually by a combination of shear welds, on a scalloped profile, along with cement grout injection. This latter is needed in any event to prevent vibration.

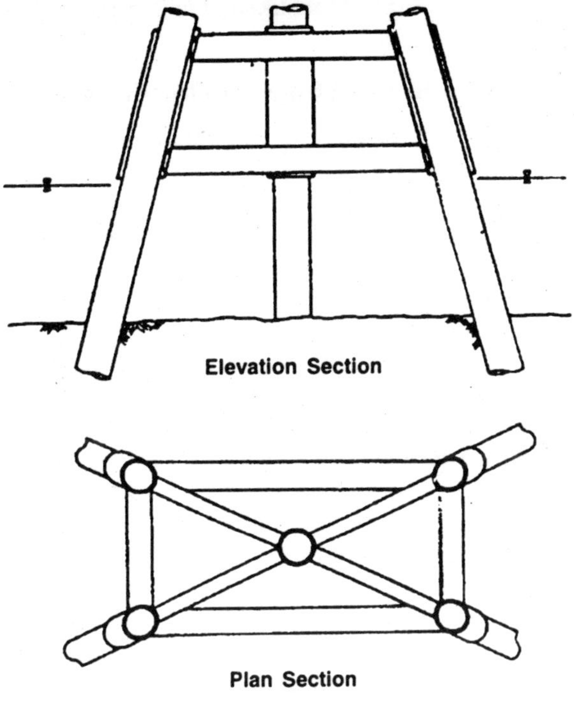


**Figure 8.3: Use of Prefabricated Template for Construction of Mooring Dolphins**

Where a jack-up construction barge is used, as has been the case for a number of offshore terminals in Japan and elsewhere, then the jack-up platform can provide the initial support for the template. However, the template should not be rigidly attached to the jack-up platform but be free to slide as dynamic lateral loads are applied by the driving of the batter piles; otherwise, the stability of the jack-up rig may be endangered.

During the construction of an offshore terminal, with its many independent structures, the derrick or construction barge will have to move many times in order to be able to handle the batter piles at the many different angles. In turn, it must have its mooring lines out at various angles, which change as the barge is moved. It is, of course, undesirable to have a line run around a dolphin, as it may apply a high lateral force in a direction other than that for which the dolphin was designed, displacing it. More likely, it will break the line just at the time it is most needed.

## The Positioning of the Mooring Dolphins

The positioning of the mooring dolphins is usually not critical; a meter or two each way is usually acceptable. The relative positioning in and out of the breasting dolphins and loading platform is very critical, because when the ship berths and lies against the breasting dolphins, it must not hit the loading platform, even with temporary deflection. At the same time, it must be close enough to allow hose connections to be made or to stay within the shiploader's radius. Hence, great care must be taken in establishing the front face for the breasting dolphins and the setback for the loading platform. Massive fenders are provided on the breasting dolphins to absorb the impact energy during docking. Typically, the 250,000-DWT tanker docks at 15 cm/s (6 in./sec), and much of this energy must be absorbed by the fenders plus elastic distortion of dolphins. These fenders therefore are large, massive, energy-absorbing devices with a predetermined load deflection response. Many different types have been developed, utilizing deforming rubber fenders, springs, hydraulic rams, the deflection of high-strength steel tubulars in bending or torsion, or the potential energy of gravity weights.

Regardless of these details, the fenders must be properly and accurately set and installed by the contractor. Fender units are prefabricated in the largest segments that can be conveniently handled. Temporary guides should be installed which will automatically position them in proper position for bolting or welding.

The previous remarks about the difficulty of welding in the splash zone apply here. Because of the impact forces involved, for example, 300 tons (3 MN), very extensive shear welds are required.

High-strength bolts are therefore usually preferable. To aid in fit-up, slotted holes should be provided, slotted in one direction on the jacket frame, the other direction on the fender bracket. If proper bearing plates are used under the bolts, a good connection can be rather quickly made. Tolerances are very important, in that the face of the fenders on the two or more breasting dolphins must line up to be engaged simultaneously and equally by the ship's hull.

The main emphasis by the constructor, therefore, must be on maximum prefabrication, provision for tolerances, and adoption of all practicable expedients to aid in installation.

The superstructure of the loading platform is now installed. To the greatest extent practicable, it should be prefabricated in large modules. The hydraulic loading arms are especially time-consuming to set, because of the accuracy required and their awkward shape. Prefabrication (pre-assembly) of these in a modular frame will save many hours of time at the site.

Similarly, erection of a shiploader on the loading platform is a major task due to the heights involved and weights that must be lifted. It may be necessary to erect a stiff-leg derrick on the loading platform in order to set the higher segments. Special planning is required for the rigging when installing a telescoping shiploader boom; not only is the boom heavy and awkward and required to be lifted high above the deck and the sea, but it must be traversed into the closely fitting housing. In turn, the lifting slings tend to foul on the housing frame and may require relocation during erection. For these reasons, complete prefabrication and assembly of a shiploader in the shipyard or harbor is, of course, preferable to assembly over water, whenever this is practicable.

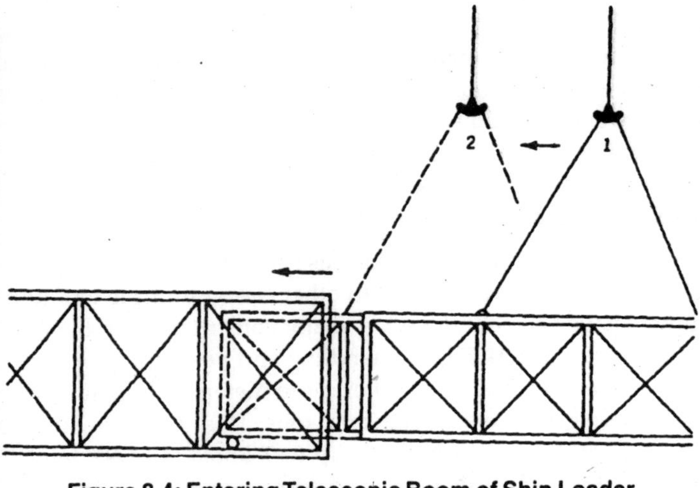


**Figure 8.4: Entering Telescopic Boom of Ship Loader**

Caisson-type (GBS) offshore terminal structures have been built in order to permit complete outfitting in a protected harbor, including shiploader, conveyor stacker and fenders.

For the Hay Point Terminal in Queensland, Australia, the terminal consisted of three berthing caissons, one carrying the shiploader completely erected, another the conveyor stacker. Short temporary buoyancy tanks were attached to the caisson bases in order to give positive stability during the critical stage of submergence below the top of the base raft.

The submergence was carried out in the harbor so the tow was performed with the structure riding with the waterline up on the shafts. These caissons were each towed out and installed in a single day, the actual set-down taking only a few hours. Mooring lines were run from each caisson to pre-set mooring buoys. Winches on the caisson pulled the structure to exact location and held it there during set down. One short mooring line was required, leading to the previously set structure. This was affixed to a rubber cushioning device that was designed to absorb shock loading. A fiber line could have been similarly used to accept the dynamic force variations.

The structures were designed to arrive at high slack water, be positioned during the fall in the tide, ballasted down to seat at low slack, then ballasted to stay on their pads during the subsequent high tide. (Tidal range was 5 to 6 m.)

Dolphin caissons had superstructures of tubular structural steel; they required substantial temporary buoyancy tanks to be attached to ensure stability and control during seating. After seating, the spaces under the caisson bases were filled with grout containing a thixotropic admixture. Scour protection, in the form of articulated concrete block mats, was placed on the periphery of the caissons.

In many cases, the installation has been carried out using large offshore-type impact hammers. In other cases, multiple vibratory hammers have been employed. A major construction problem for this type of construction is the handling and positioning of such a large-cylinder pile, say 50 to 60 or more meters in length, 3 m in diameter, weighing several hundred tons. Tidal currents, such as those in Cook Inlet, which range up to 7 knots (4 m/s), tend to displace the cylinder pile both in the direction of the current and laterally due to vortex shedding. The pile must have special slings fitted so that it will hang vertically. Once set into the soil, the problems of current are reduced. The next problem is that of driving such a large pile in the heterogeneous soils which are typically encountered: glacial till and overconsolidated silt in Alaska, caprock, limestone strata, and calcareous sands in the Middle East areas. The subsequent superstructure erection is carried out in similar manner to that for the more conventional terminals, except that prefabrication is made easier by the all-vertical pile arrangement. Alternatively, very large prefabricated deck sections or bridges may be set by a crane barge. These may incorporate piping and equipment.

The Arco Terminal at Cherry Point, Washington will employ 2-m-diameter steel tubular vertical files, driven to 25-m penetration by means of a very large hydraulically actuated drop hammer. Smaller-diameter stub files will be concreted into the large-diameter piles to give greater flexibility and ductility under severe earthquake. Prefabricated deck sections will be completely outfitted to reduce the subsequent topside work to essentially that of connections.

On some offshore terminals, for example, the LPG terminal at Ju'Aymah and the petrochemical terminal at Jubail, both in Saudi Arabia, large-diameter prestressed concrete and steel cylinder piles have been set in predrilled holes and then driven to develop proper bearing and lateral support. These have varied from 1.6 to 4 m in diameter.

Trestle connections from shore are relatively standard in their construction operations. Because they typically cross through the surf zone and over shallow-water areas, part or all of their construction is carried out over the top. By such methods, the lifting, driving, drilling, and framing are essentially independent of the sea state and current. Typically, the upper works of a large crawler crane are mounted on long girders, for example, double wide-flange beams designed to span one bay and having extensions that cantilever out to the next bay.

Spud piles are dropped and the girders are jacked to grade. The crane then sets the piles through the template and drives them. The template is now welded to the piles. Longitudinal stay beams are dropped into place and bolted. The rig can now skid forward to the next bay. Another crane follows up behind, placing prefabricated deck sections, completing all bracing and framing. The piles in the template need to be fully fixed to prevent excessive vibration and assure proper interaction. Shims are inserted to fix the pile head,

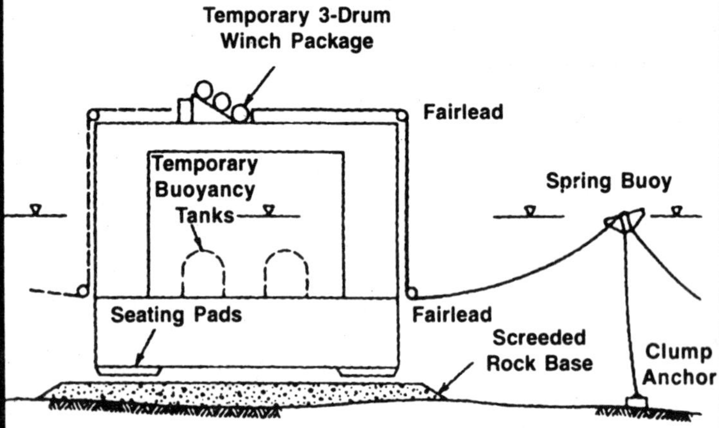


**Figure 8.5: Positioning of Offshore Terminal Caisson**

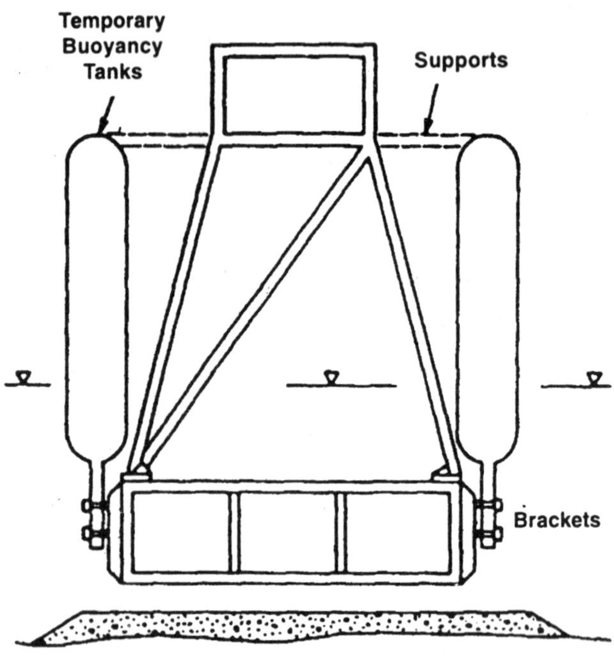


**Figure 8.6: Use of Temporary Buoyancy Tanks to Provide Stability and Control During Installation**

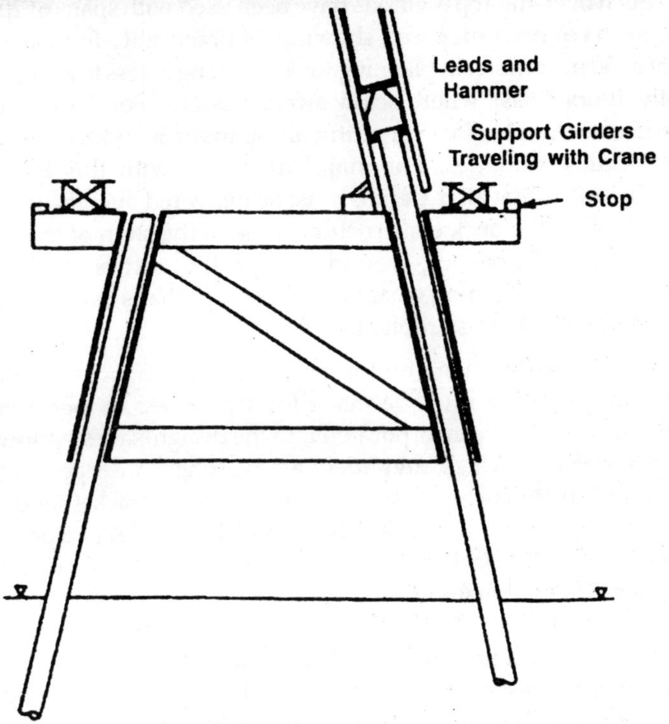


**Figure 8.7: Template and Bent Frame for Piled Access Trestle to Offshore Terminal**

and a seal made at the lower end of the sleeve, followed by grout injection and welding at the top. Accelerators may be used in the grout to achieve early strength. A long trestle will require an anchor bent every 300 to 500 m; this should consist of a double-bent, adequately braced. Because of weight limitations, the prefabricated template may have to be set in two half-segments and bolted together.

Launching forward of the rig is done by pulling against the longitudinal girders. To guard against accidental sideways displacement, lateral stops should be installed on the end of the cap girders of each template. Such a sidewise runoff actually occurred when launching the crane rig forward on the access trestle to the Hay Point terminal, Queensland, Australia; fortunately, the rig did not roll off, and no damage or injury occurred.

Such over-the-top methods have been used with spans of 20 m, designs have been prepared showing its practicality for spans as great as 30 m or more. In planning for a very long access to a terminal off the Ivory Coast, where heavy swells from the Southern Ocean would make work afloat very difficult, spans of 30 to 40 m showed significant economies. The major difficulty with this type of construction is that all materials must be delivered over the trestle, rehandled by the deck construction crane to the stern of the pile-driving rig, then swung around by it to the next bent. Proper packaging and planning of such deliveries, perhaps using a night shift, will simplify this problem.

Even when working on top of a trestle or offshore terminal platform, provision must be made for safe access by personnel. Walkways need to be incorporated into the design, safe stepdowns when the crane swings, and adequate lighting. In deeper water, trestle supports may be constructed afloat, as small jackets and pin piles. Water safety also needs special consideration. A person may fall overboard. Life jackets should be mandatory. Lifelines of nylon or similar floating material should be strung and a rescue boat on call or patrolling as indicated. From a modern terminal, with deck at +10 or +12 m, and nothing but tubular piles in the water, it can be very difficult to rescue a person who has fallen overboard in a choppy sea with a current running. The person needs to have a line to hang onto while awaiting rescue.

Many of these existing structures have adequate vertical carrying capacity but require stiffening to prevent excessive drift, with its amplification of moments by the P-Delta effect. Incorporation of large-diameter vertical steel cylinder piles, tied into the existing structure, has proved an effective retrofit in many cases.

Shear resistance of existing pier footings and shafts can be improved by banding with heavy steel plates. These are installed around existing elements, joined by bolting, and grouted to fill the annulus. Where they can be joined above water, and then slipped down, welding is employed.

Wharves and terminals with intersecting batter piles or batter and vertical piles have suffered severe damage in earthquakes. Examples include the wharf at Anchorage, Alaska, in which the steel tubular batter piles were severely distorted into S-shapes and punched through the deck, and the many wharves in San Francisco

Bay where the concrete batter piles were crushed and sheared during the Loma Prieta earthquake. Similar but lesser effects occurred during the Northridge earthquake.

Retrofit has consisted of cutting loose the batter piles, replacing them with vertical piles, and then extending the wharf landward so that new vertical anchor piles (shear piles) could be incorporated. These are typically large-diameter piles with special reinforcement and confinement to take the concentrated moments and shears at their heads.

*Chapter 9*

# The Removal of Offshore Structures and Construction of Offshore Platforms

## Introduction

A number of studies are currently under way, to reappraise the regulations and to re-evaluate the need for removal to below the mudline. Current regulations of many countries require the removal of offshore platforms and other structures, when they have finished serving their purpose and are no longer in use.

Three types of structures will be addressed in order to illustrate general principles and possible solutions for removal. Obviously, any particular platform will have to be addressed in specific detail as to requirements, methods, and control of operations.

Removals are in many ways as complex as or more complex than critical installation. The existing conditions must be fully investigated and considered since structures may be corroded, damaged, or even have missing braces. Marine growth and seafloor changes have to be considered. Planning of each stage must be carried out with thorough attention to detail. Whereas in initial installation there are economic incentives for the owner, the dismantling is a net economic cost.

Perhaps even more than in initial installation, there are risks involved in removal: risks that during salvage an accident will occur or the structure become unstable, presenting the constructor with a more difficult and costly or even a nearly impossible operation. Therefore, just as in installation, risks must be enumerated and evaluated and contingency plans prepared.

## Removal of Gravity-base Platform

Gravity-base platforms have primarily been built of reinforced and prestressed concrete, although several are of steel and some recently constructed GBS platforms are of hybrid steel-concrete construction.

These platforms are characterized by large base caissons, the latter having originally provided flotation during transport and installation. A major concern in removal is that of excessive breakout resistance due to the increases in soil shear strength, cohesion and adhesion of soil with the base as well as with skirts and dowels.

The basic concept in salvage and removal is to refloat the structure and then tow it to a disposal site in deep water. A typical procedure might be as follows:

1. Cement and plug wells. Purge and disconnect connecting flow lines and risers.
2. Remove conductors and attached well casing. Plug conductor holes in base with concrete plugs. Cut piping connections loose and plug all penetration by concreting.
3. Remove equipment and facilities piecemeal from deck, cutting deck frame into sections as required for removal by crane barge. Alternatively, the deck structure can be removed as a whole. Preferably its weight has first been reduced by removal of selected items. The deck is cut free from the shafts. Then the same barge arrangement used to install the deck, typically three barges, can be moved in under the deck so that by deballasting, the deck can be removed and transported to an inshore site for further dismantling.

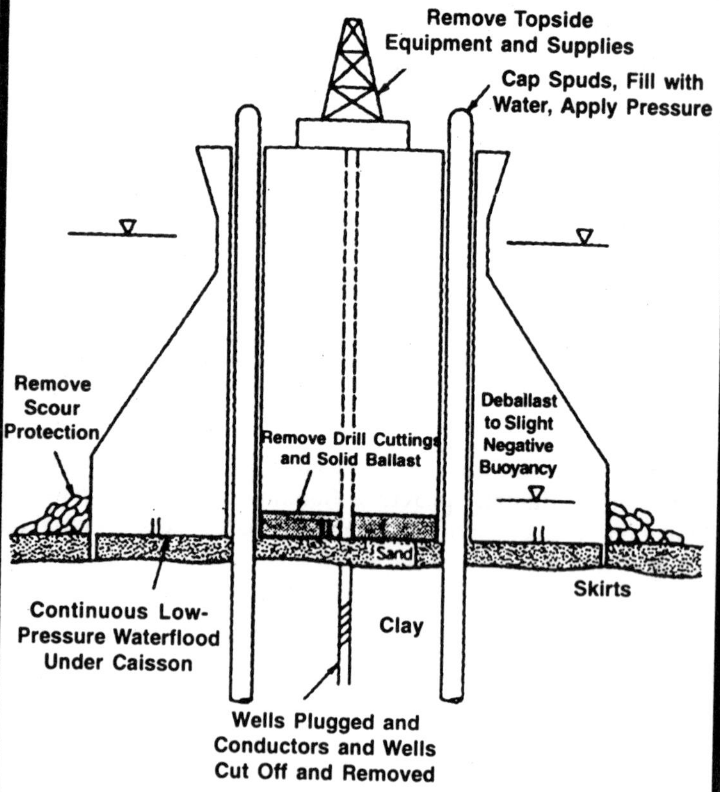


**Figure 9.1: Removal of Gravity-base Platform**

However, the initial installation was performed under quiet, protected conditions whereas now, the work is being carried out in the open sea. Typically, clearances between barges and shafts were of the order of 100 to 150 mm which are impracticable in the open sea. Further, as the barges are deballasted to make contact, they will be heaving under the swells, no matter how calm the surface. Hence, special hydraulic jacks and compressible contact devices will be needed to cushion the impact.

1. Salvage equipment from riser and utility shafts.
2. Remove exterior ballast walls near base (cut post-tensioning ties), allowing ballast to spill out over seafloor.

Plug any openings which may have been formed or cut for piping penetrations.

3. Remove solid ballast in interior that may have been placed after installation on site. This ballast will probably have been either sand or slurried iron ore. This can be removed by airlift or eductor or specialized equipment such as Toyo submersible pumps, which have jets incorporated to agitate the material and slurry it, facilitating its removal by airlift or eductor. The extent of removal depends on the computation of weights and availability of access to compartments. It would, of course, be desirable to install access sleeves or manholes in the platform at the time of original construction to facilitate this operation.
4. At this stage, the ballast compartments are fully flooded. Using pipes leading to the underbase, inject water underneath at a low, steadily maintained pressure, slightly above ambient at the base elevation. The pressure must be low enough that it cannot cause piping under the skirts. (Once piping occurs, little additional benefit can be attained by underbase waterflood.) Maintain pressure for up to 24 hours.
5. Deballast caisson to a slight positive buoyancy. If structure does not break free, deballast one side more than the other to tip caisson off. Once caisson breaks free at one edge, water will be sucked in underneath and break all suction. Of course, if there are no skirts, water can flow under the base freely once the edge lifts off.

   Limit deballasting to the point where, if structure breaks free, it will not rise above the level at which it is still fully stable. This can be a very critical stage because some excess positive buoyancy must be provided to extract the dowels. Once the structure breaks free, it will rise until equilibrium; is reached. A special check must be made to ensure that at this stage the structure is still stable.
6. Tow structure to disposal site.

Deballasting should normally be by pumping. The use of compressed air is generally not desirable and may be dangerous. First, if used under the caisson to help overcome suction, the high-pressure air will tend to escape laterally, leading to piping. Second, if compressed air is used internally to expel the water, as the structure rises, the external head decreases. The air then expands more. Thus the structure tends to rise farther and faster than planned. Third, this expanding bubble of air creates a free-surface effect, traveling to, the high side, where it exerts even more upward force, developing an overturning moment.

In ship salvage, the use of compressed air has led to disastrous results when excess air is pumped in to overcome the suction. The ship breaks loose, and rises up. The air bubble expands. The ship's rise accelerates. It rises to the surface, but now the air bubble is all on one side and the ship turns over. The air escapes, and the ship plunges back to the bottom!

It is obvious that extremely careful calculations are needed, taking into account not only the weight of the original structure and its displaced volumes, but also changes that have occurred since, such as:

1. Marine growth.
2. Drill cuttings or sediments stored inside.
3. Weight of underbase grout which sticks to base.
4. Setup of soil on skirts and dowels, increasing extraction force.
5. Ballast or dropped material on caisson roof.

In addition, the structural adequacy of various critical members must be looked at carefully, since the uplift forces exerted by buoyancy are extremely great. Physical damage may have occurred over the intervening years.

## Construction of Offshore Platforms

So far we have learnt about the techniques of removal of offshore structures. Now let us try to understand about construction of offshore platforms.

The construction of offshore platforms has been a heavily cyclic industry, responding almost frantically to the discovery of new oil provinces such as the North Sea or significant changes in price level, such as that which followed the OPEC oil embargo of 1973–74. The discoveries on the Alaskan North Slope and Canadian

Beaufort Sea triggered a major effort in the Arctic, with its new environmental loadings from sea ice and icebergs. Then there was a hiatus of almost 10 years, during which the market stabilized, the industry matured and became more orderly in terms of construction and cost. By the time the current boom (1997–98) arrived, technology had also changed, requiring a learning curve before the industry as a whole was geared to the new demands. The current boom offshore focuses largely on deep water while, at the same time, the reduction of the cost and size of platforms and the increasing movement to floating production systems.

The result has been that in the periods of upward surges in demand, there have been very significant cost overruns and schedule delays in the early periods of each new development, followed by a gradual steadying as estimates rose to meet actual data and competitive forces brought costs down closer to their targets.

The cost of structures has a significant influence on the viability of the offshore development, largely because it is an early capital expenditure. Similarly, the time lag between structure expenditure and oil income has tended to increase rather dramatically as the

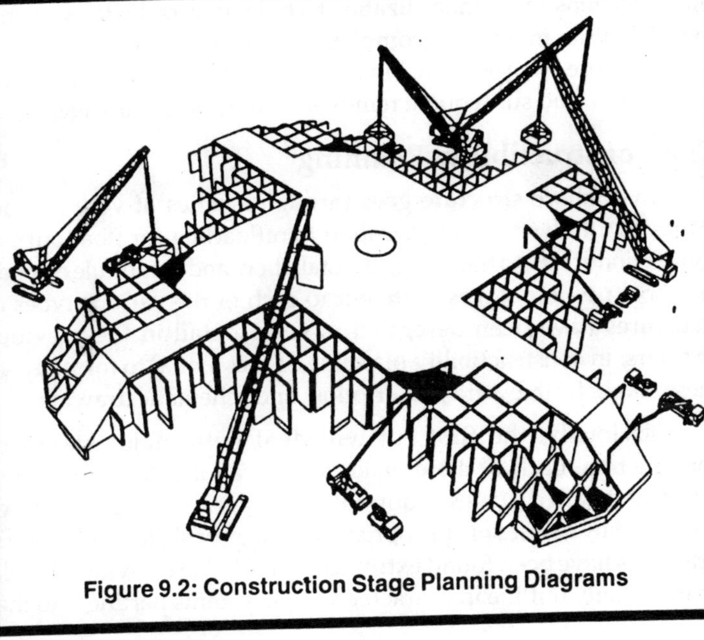

**Figure 9.2: Construction Stage Planning Diagrams**

projects have moved to environmentally more demanding areas–deeper water, more remote locations–and with greater ecological, social, and political constraints.

Nevertheless, the offshore structures per se are not usually the major cost of field development. However, they are a significant part of that total cost, and they are usually on or close to the critical path in the schedule. They comprise one area where sound innovative design, competent construction planning, and competent construction management can achieve meaningful savings.

Constructibility is involved in the concept development and the integration of design with construction. It determines the selection of construction methods, facilities and stages, the procurement and assembly of materials and fabricated components, the organization and supervision of the work, and the training of workers. It includes analysis and planning, quality control and assurance, safety engineering, cost estimating, and budget control.

It also includes an item of special concern to offshore structures: weight control. It addresses personnel and material transport and access, craneage and the planning of heavy lifts, and the furnishing of utilities during construction. Constructibility employs work simplifications and standardization techniques in order to overcome the difficulties inherent in complex and sophisticated construction in an offshore environment. Finally, its scope includes deployment, installation, and subsequent removal, relocation, or salvage.

## The Constructibility Planning

An offshore structure goes through a series of very distinct stages as it moves from fabrication to offloading (or float-out), to completion afloat, to transport, to installation, and to module erection and hookup. The stages pertinent to each of the various types of structures have been described in some detail in the previous chapters. In constructibility planning, it is essential to formally set these stages forth by title, description, and schematic drawing.

Obviously, the first cut will deal with major stages of construction. Each of these major stages can then be subdivided into the detailed stages required. The stages should be further portrayed by a series of appropriate drawings or sketches. Isometric drawings have been found extremely useful. The drawings should be essentially outline in character, with key items pertinent to that

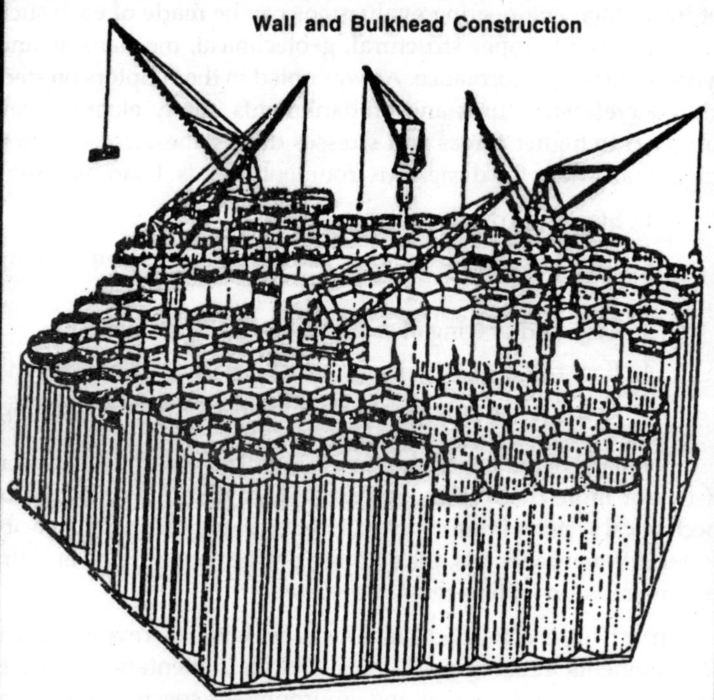


**Figure 9.3: CAD Applied to Constructibility Planning**

stage shown in heavy lines. The purpose is to eliminate aspects not essential to that stage so that the key elements can be clearly recognized. Thus, while they are based on engineering design drawings, they differ from them in emphasis, clarity, and use. Computer-aided design (CAD) is especially effective in enabling three-dimensional portrayal of the successive stages.

Experience in the preparation of such descriptions and drawings has shown that serious errors have occurred due to "jumping past" intermediate stages, which have been incorrectly assumed to be unimportant or self-evident. The entire purpose of constructibility planning is negated when this happens, because it is just these skipped stages that so often turn out to be critical.

Once the constructor is satisfied that all the stages have been set forth, then engineering evaluations can be made of each such stage to ensure proper structural, geotechnical, mechanical, and hydrodynamic performance. As was noted in the chapters on steel and concrete structures and embankments, many elements are subjected to higher forces and stresses during these construction stages than under the design environmental loads. Examples are:

1. Steel piles during driving.
2. Pipeline bending and radial compression during installation.
3. Legs and bracing of steel jackets during launching.
4. Base raft of gravity-based structures during float-out.
5. Cell walls of gravity-based structures during deck mating.

For many of the stages, the key issue will involve the interaction of two or more disciplines. For example, ballasting by means of mechanical systems is intimately connected to the structural capacity under differential heads, the stability performance afloat, and the instrumentation with its real-time readout.

In offshore construction, however, with its revolutionary developments in equipment, tools, and instrumentation, with its new structures and systems and environments specific experience may not exist. Instead of relying solely on intuition, therefore, the conscious use of constructibility planning and evaluation of stages should lead to a more rational and effective program.

Some of the principles which can be beneficially applied to reduce the time and cost of construction are:

1. Subdivision into as large components and modules as is possible for fabrication and assembly.
2. Concurrent fabrication of major components in the most favorable location and under the most favorable conditions applicable to each component.
3. Planning the flow of components to their assembly site.
4. Providing adequate facilities and equipment for assembly–the fabrication site must have adequate space for subassembly, storage, and access; the special equipment may include such items as synchrolifts, heavy-

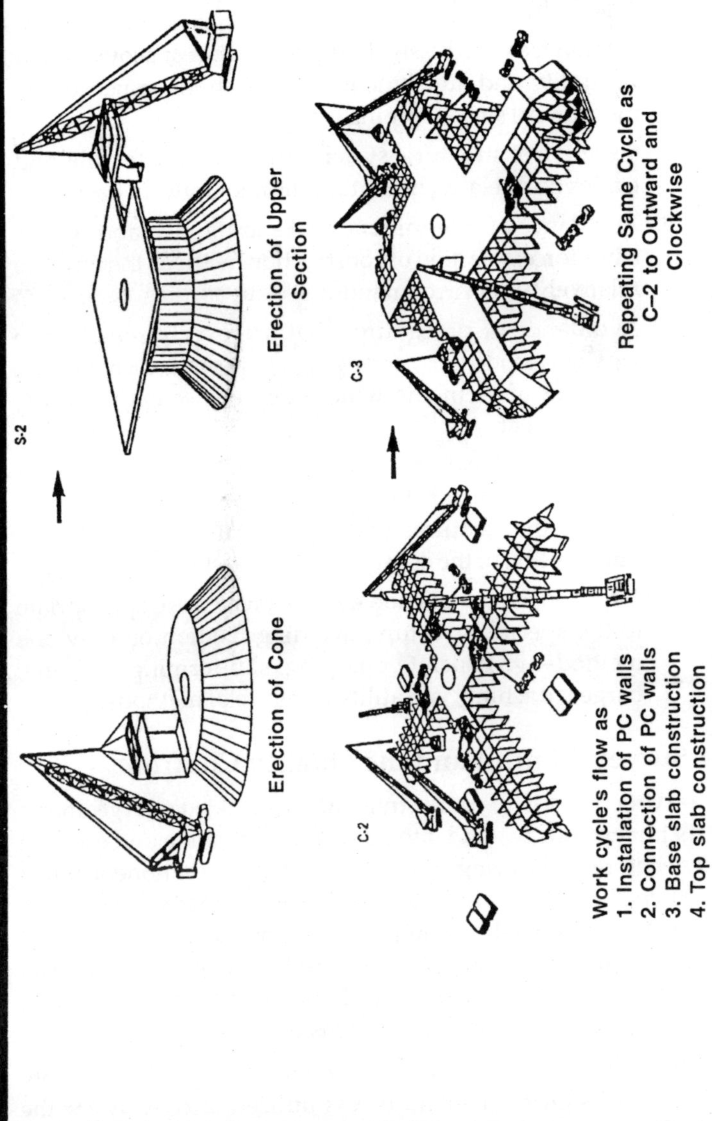


**Figure 9.4: Stages of Construction Portrayed by CAD Schematics**

lift cranes, both land-based and barge-mounted, dry docks, and construction basins.

5. Simplification of configurations.
6. Standardization of details, grades, and sizes insofar as practicable.
7. Avoidance of excessively tight tolerances; provision for flexibility and adjustment in connections, especially in mechanical system piping.
8. Selection of structural systems that will utilize skills and trades on a relatively continuous and uniform basis.
9. Avoidance of intermittent peaks in the demand for the labor force; selection of construction methods that involve relatively uniform demand.
10. Avoidance of procedures that are overly sensitive to weather conditions; arranging shop prefabrication and painting of elements which are very sensitive to the environment.
11. Modularization of mechanical systems to be incorporated in or on the structure into the largest possible components, even if this requires additional structural support or interruption of the construction of the structure proper.
12. Selection of construction methods which are appropriate to the specific structure, avoiding fixation on only one method–*e.g.*, concrete pumping, slip-forming, welding, barge launching; versatility in choice of methods.

## The Survey Control and the Quality Control

Survey control is, of course, intimately associated with geometry control but extends beyond it to guide the fabrication and erection process before and during construction. Where components must fit to others, it is often difficult to establish the proper reference line. One must be selected–for example, a line connecting the center of two best-fit circles. The other points must be properly related in all three planes. Templates will often be found to be the best method of transferring complex interactive dimensions.

Match-casting of precast concrete members has been highly successful in assuring later fit. It was utilized effectively for the breakwater segments of the Ninian central platform. Care has to be taken in such match-casting to avoid distortions due to thermal effects, for example, warping during steam-curing. Match-casting, followed by erection with epoxy-glued joints and post-tensioning has been widely employed in bridge construction.

Similar match-fitting and templating can be used with steel fabrications, again recognizing potential distortions due to welding. Templating has been effectively used by the Japanese module fabricators to ensure proper fit between adjacent modules and thus facilitate connection.

Proper survey control procedures must also be set up for erection of space frames, of which the steel jacket is the most common example. Distances are large, 50 to 100 m or more, and points are high in the air and hence of limited accessibility. The sun's heat may cause significant elongation of upper members during the afternoon, while lower members on the ground are partially restrained by friction as well as perhaps seeing lesser temperature rise due to shade. Deadweight deflections may account for even greater distortions, since the jacket is usually fabricated in a different attitude from that in which it is installed. Diagonal measurements often provide the best check.

When structures are afloat, there is always difficulty in establishing reference lines, especially the vertical lines. Lasers can be rigidly mounted at the base, accurately set normal to the base. They can then project a normal line, called "vertical," even if the structure is slightly listed due to ballasting or deadweight.

Even a relatively rigid structure such as a concrete gravity-base structure undergoes significant deflections during construction due to deadweight and ballasting eccentricities. Thus the shafts may deflect outward during ballasting down to receive the deck.

The establishment of a quality control (QC) manual and a quality assurance (QA) program is an essential aspect of constructibility. The first task is to set up what the requirements are. They, of course, include those specified by the designer. If the designer has used only general requirements, such as "compliance with the specifications," it is necessary to determine which elements of that specification are important and, further, which will be determined or measured in the construction process. To these constructors must add those requirements necessary to enable them to carry out their work in accordance with the materials selected and procedures adopted–for example, temperature and humidity control for painting; moisture control for above-water embankments, and early strength for concrete.

In establishing these lists, every effort should be made to reduce the number to the bare essential minimum. The nuclear reactor syndrome that "everything that can be measured must be measured" must be avoided. Paperwork must not become more important than the structure. The quality assurance program should then provide for the identification and recording of the critical items which may be important for future reference. QA should not be used as a whip by which to ensure that the inspectors are doing their job.

Where defects can be immediately corrected, they should be.

Agreement should be reached before construction starts regarding what records are to be recorded and which data (*e.g.*, radiographs) are to be kept. Those which are so kept must be properly identified and stored. Only that which is essential for the proper performance of the structure should be tested and inspected. Only that number of tests should be made which are necessary to ensure maintenance of quality on a statistically defensible basis.

The reason for the above exhortations on limiting inspection and tests is that experience has shown that usually far more data has been collected than can be reduced, evaluated, and used, and that by so doing, insufficient emphasis is given to the key properties which are truly important for performance.

An offshore structure is a major undertaking on two fronts, because of (1) the effect of the sheer size, complexity, and interdisciplinary aspects, and (2) the dynamic movement, transport, launching, upending, and submergence that must be carried out on a grand scale, sometimes involving over half a million tons and a structure the size of our large high-rise buildings.

Construction management will have carefully planned each operation. Now as the work goes on, how is the success or lack of it monitored? What warning signals will be sent, and how will they be recognized in time for corrective action.

Referring first to the productivity of fabrication and erection, careful monitoring can be carried out on the basis of schedule, unit costs or percentage of completion, man-hour or crew-day requirements, all compared with budgeted costs and time. It is not enough to try to control by flagging exceptions; the 10 per cent overrun or underrun may apply to an insignificant item or one which will soon be completed and hence beyond timely correction.

Rather, the major components of the work need to be identified: schedules and budgets assigned, with consideration of the learning curve, and the special conditions. These key items are then closely monitored, usually on a crew-day basis.

Constructibility planning must, of course, include an interface with the critical path scheduling. The critical path method (CPM) is a valuable technique for evaluating and controlling the various operations. The growing use of microcomputers in the field enhances the ability to identify critical elements of progress early, enabling appropriate action to be taken.

Critical path schedules are, of course, constantly updated. While most attention goes quite naturally to the items that lag, consideration must also be given to the opportunities that present themselves when work goes faster than scheduled. The Statfjord C platform wa head of schedule on several early items; others were then accele ed to enable the completed structure to be placed on station several months ahead of target.

The second type of construction control relates to technically critical operations. What early indications will there be if serious engineering problems are imminent? Prior study has to be given to this matter for each critical operation. The instrumentation can be installed and observation schedule and procedures established to ensure that timely warning is received.

Examples of early feedback are unexplained discrepancies between weight control, ballast control, and observed draft. Another example is a trim or list that is inexplicable or beyond predictions. Rupture of erection bolts may indicate excessive built-in stresses. Cracking of welds may be due to poor welding or to excessive stress. Residual stress has been identified as the primary factor in the weld cracking in buildings which have been subject to earthquake. Cracks in reinforced concrete may be local and caused by shrinking or may indicate major internal delamination.

In the upending process, is the attitude matching that predicted on the basis of ballasting calculations? If not, watertight closures may have ruptured, valves may be stuck open, or piles that were carried may have broken loose. From detailed consideration of each major observation, the needed data, their timeliness, and their relevance can be determined.

Experience on major projects onshore and offshore where serious accidents have occurred has shown in hindsight that warning phenomena had often been observed but had been disregarded because of overconfidence that the engineering and construction control was infallible.

*Chapter 10*

# The Technique of Concreting and Grouting

## Introduction

In the construction of offshore structures, underwater concreting and grouting, play an important role. Underwater concrete may be placed in forms to serve as the structure itself, most often it becomes the footing block for the structures. It may be used to tie together various elements in composite action–for example, to tie the piling to the footing and thence to the upper portions of the structure. Underwater concrete may also be used to fill pre-excavated holes in the seafloor and to act as a leveling met. Underwater concrete may also be used as solid ballast to add weight and lower the center of gravity. It may be used to fill under the base of a gravity platform to insure uniform bearing and provide shear transfer. Underwater concrete may be placed in piles or caissons to give added structural strength and to prevent buckling, or it may be placed in belled footings which have been drilled at the tip of the piles in order to increase axial compression and tension capacity.

Underwater grout is also used for many of the above purposes; hence, no rigorous distinctions should be made in classification. Grout is also used to bond piling to jacket legs, to cement well casings, to fill small spaces between elements to provide structural continuity, and to fill the voids in preplaced rock and aggregate.

Underwater concrete should be proportioned to develop a plastic, highly workable, and cohesive mix, not subject to segregation.

For many structural purposes, the following mix is suitable:

**Coarse Aggregate**

Gravel of 20 mm (¾ in.) maximum size. Use 50 to 55 per cent of the total aggregate by weight. For congested areas, use 10 mm maximum size aggregate (pea gravel).

**Fine Aggregate**

Sand, 45 to 50 per cent of the total aggregate by weight.

**Cement**

Type II; ASTM, 350 kg/m$^3$ (600 lb/yd$^3$).

**Pozzolan**

ASTM 616 Type N, F, or C: 60kg/M$^3$ (100 lb/yd$^3$).

**Water-Cementitious Material Ratio**

(w/cm), 0.42.

**Water-Reducing Admixture (preferably it is also a plasticizer)**

Do not use super plasticizers without a retarder.

**Retarding Admixture**

To increase setting time to 6–24 hours, as required.

**Slump**

160 mm ± 15 mm.

This mix will develop compressive strengths in the range of 40 MPa at 28 days [(5600 psi) (cylinder strength)]. It will generally flow out on a slope of 6:1 to 8:1 and, if properly placed, should give minimal segregation and laitance. It is suitable for placement in voids as small as 300 mm in diameter and can be used for large caissons and bridge piers.

For higher strength and greater cohesiveness, silica fume may be added in the proportion of 5 to 6 per cent by weight of cement. However, in this case, it is necessary to use superplasticizer, either alone or in combination with conventional water-reducing admixture. Adequate retarder must be added to ensure that the time of initial set is long enough (typically 6 hours) so that premature stiffening will not occur. Where it is important to obtain a level surface and minimal or no laitance, anti-washout admixture (AWA) should be added. For very large pours, larger-size coarse aggregate (40 mm) has been used but shows little economy because of the need for the large proportion of sand. For smaller pours, the coarse aggregate may be reduced in size, to as small as 10 mm (3/8 in.).

The basic mix recommended above, however, will develop a fairly high heat of hydration (about 35°C above ambient) depending on the size of the placement, leading to thermal expansion and possible cracking during the subsequent cooling. Various methods of reducing the temperature rise are available. Their use is justified in special cases. The following is a list of individual means which have been employed (they should not necessarily be combined):

1. Select aggregates with a high thermal coefficient, requiring more heat per degree of temperature rise.
2. Use blast furnace slag/cement in the proportion 70:30 to reduce heat generation. Slag should be coarse-grind (< 3800 $cm^3/g$), with no added gypsum.
3. Increase the percentage of pozzolan (fly ash), replacing a comparable portion of cement. Recent tests indicate the percentage may be increased to as much as 50 per cent in fully submerged unreinforced concrete.
4. Use limestone powder to replace part of the cement.
5. Precool the aggregates by water spray-evaporation and use ice as the mixing water.
6. Cool the mix by injection of liquid nitrogen.
7. Subdivide the pour to reduce the size of individual blocks.

Since various admixtures, such as pozzolans, behave differently with different cements and admixtures, trial batches of several cubic meters should be made to ensure that the resultant mix is workable and possesses a high degree of cohesiveness, that is, does not tend to segregate.

## Underwater Concreting

The placement of tremie concrete is carried out through a tube. Usually, the tube must be at least 8 to 10 times the maximum size of coarse aggregate. For typical placements, 200 to 300 mm diameter tubes are usual. The pipe may be sectional, but joints should be flanged and bolted, with a soft rubber gasket or screwed joints to prevent any in-leakage of water. When the mix is poured down the pipe, if there is a gap in the joint, there will be a venturi effect which will suck in seawater and mix it into the concrete thus grossly increasing segregation and washout. Where practical, placing the tremie pipe on a slant of 5° to 10° from the vertical will allow entrapped air to escape.

The preferred way to start a pour of any depth up to 50 m is to install a steel plate on the bottom end with a soft rubber gasket. The plate is tied with twine to the pipe. The tremie pipe must have sufficient wall thickness so that it is negatively buoyant when empty. The pour is started by placing the sealed pipe on the bottom and then partially filling it with the, tremie concrete mix. While there will be some segregation during the fall down the empty pipe, there will normally be adequate remixing at the bottom. To ensure this first, place 1 $m^3$ of the mix with the coarse aggregate omitted.

When the tremie has been filled to a reasonable distance above the balancing head of fresh concrete vs. seawater (about 50 per cent), the pipe is raised about 150 mm, allowing the concrete to flow out. "Reasonable distance" is that required to overcome the friction head, which may be only a meter or two. The lower end of the pipe is kept embedded in fresh concrete, but always above the level where the concrete has taken initial set. With a retarder to prevent initial set, the depth of embedment becomes less sensitive. The tip of the tremie pipe should be immersed about 1 m as a minimum to prevent water inflow into the pipe. The flow of concrete should be smooth, consistent with the rate at which concrete can be delivered into the hopper at the top. Similarly, the method of delivery should provide relatively even feed into the hopper rather than large batches being suddenly dumped. If for any reason the seal is lost, the concrete having flowed completely out of the pipe, then the pipe should be raised and sealed and the pour re-started as in the original program.

When large areas are to be covered, multiple tremie pipes should be used or the tremie pipes reset in new locations within the slope of the fresh pour. The distance tremie concrete can flow without excessive segregation is between 6 and 20 m: the larger distances are obtainable with a flowing but very cohesive mix, which prevents excessive segregation and washing of the cement. The slope of an extensive pour will allow the laitance and silt to flow down so that it tends to collect in a far corner, where it might be trapped under good concrete. An airlift may be operated in the corner of an enclosure to remove any laitance.

The major problems with treme concrete have to do with segregation into sand, gravel, and a mixture of cement and water known as laitance, from the French word for "milk". Actually laitance may be a very plastic, claylike substance and will eventually harden into a chalk-like material. It is very porous and constitutes a weak layer in the structure unless properly removed. Segregation occurs primarily when concrete is allowed to flow through seawater or where seawater is mixed into the concrete, as by mechanical disturbance, divers walking in it, attempts to vibrate it, or use of a nonplastic mix that tends to build up and then break through in an overflow. Churning of the tremie ripe to promote flow and moving the pipe horizontally through fresh concrete are very bad practices. Leaky joints will act as a venturi mixer and wash the concrete during its downward flow. When filling under an upper portion of the structure or where high-quality concrete is needed right-at the top, tremie concrete should be overflowed to displace any laitance until good concrete appears.

A very widespread practice has been carried forward from earlier years; this is the use of a "go-devil" or traveling plug to start the pour. For example, in the earlier days, a plug of hay or burlap was placed in the pipe, and the concrete poured on top of it, forcing it down and forcing the water out of the pipe ahead of the plug. This practice, while crude, was a moderately effective means for the initial start. It should never be used when restarting a pour after loss of seal or when resetting a pipe into fresh concrete. The water forced down ahead of the plug will be forced to flow through the fresh concrete, washing out the cement. Even in starting a pour, if the bottom is very soft or sandy, this flow of water may cause erosion.

In recent years the plug of hay has often been replaced by a ball, usually a volleyball. Such a ball fits nicely in the pipe and is readily forced down by the concrete. A ball is usually inflated to only 11 psi (80 kPa). This corresponds to a depth of about 8 m (25 ft). Beyond that depth, the ball will collapse. If it later comes back to the surface, it may have re-inflated, thus disguising the problem. A ball does not provide a suitable answer except in very shallow water.

A traveling plug or "pig" is an effective and safe method by which to start a placement. It should be of constant dimensions, gasketed so that it loosely wipes the side, and self-buoyant. A wooden cylinder, steel pipe cylinder filled with foam, and polyurethane cylinder (of suitable strength to resist the hydrostatic head) are all acceptable. Rubber wiping gaskets can be installed on the pig. Such a pig may not rise back to the surface–that is, it may become trapped–but in the usual case that is of no concern. The traveling plug is thus similar to a pipeline pig. It is necessary to use such a device when very deep pours are undertaken, since over 30 m depth the use of the plate seal becomes unsuitable.

Foot valves, mechanically and hydraulically operated, have been tried for many years. They have almost always been a failure due to jamming of the coarse aggregate in the valve or setting of the initial cement grout in the valve mechanism. However, recently developed hydraulic valves which open to give a smooth bore have reportedly been used successfully in Sweden and Japan. These have the advantage that the placement can be shut off at the tip of the pipe as the pipe is raised, thus eliminating the typical mound that forms when a conventional tremie pipe is raised at the conclusion of the pour. Their use should be restricted to special cases and only where proven suitable by tests.

For placement in shallow lifts exposed to water and for placement in slowly flowing water, *i.e.*, current or wave action, one of two special solutions can be adopted. Silica fume, in an amount 6 per cent ± of the weight of cement, can be added, or alternatively, an antiwashout admixture can be added.

In both cases it is necessary to add a superplasticizer (high-range water reducer), plus a retarder (unless the superplasticizer already contains an adequate retarder).

The resultant mix, having a W/CM of about 0.45 and a slump of 250 mm, will be almost self-leveling. Either mix, silica fume or AWA, reduces the bleed. Always make a trial batch of 1 to 3 $m^3$ to verify compatibility of components and to ensure workability. The placement procedure should be the same as that prescribed in this section. In no case should the resultant mix be allowed to intentionally fall through open water, despite the claims of the manufacturers.

These same two additives, *i.e.*, AWA admixture or silica fume, can be employed for placement of mass concrete on the seafloor, using a large bucket with a closed top. AWA admixture has been used in very shallow water for filling the voids in riprap and in cyclopean construction, where alternate courses of large rock and tremie concrete are successively placed.

When using AWA, grout or concrete with small aggregate (*e.g.* 8 mm) can be placed through a hose, which is guided by a diver.

In a notable case, a mixture containing blast furnace slag cement (ratio 70 to 30) plus antiwashout admixture, was placed down a 3 in. (75 mm) pipe slanting at 7° from vertical, to a depth of 250 m. Maximum size of coarse aggregate was 10 mm. Tests showed no segregation and the resultant concrete had a compressive strength of 45 MPa and a tensile strength of 7 MPa. The water temperature at the site (northwest shelf of Australia) was 38°C, so the mix was precooled with liquid nitrogen. The pipe was precooled with cold water before starting the placement.

## Grouting

In the method of grout-intruded aggregate, coarse aggregate devoid of fines smaller than 15 mm is placed within confining walls. Grout pipes are embedded at regular intervals, horizontally and vertically. The exposed surface is covered with a mat or with an extra thickness of rock. A special grout is then pumped through the pipes to fill the interstices between the aggregate. This grout must have excellent flow characteristics and minimal bleed yet retain its general cohesiveness. Various proprietary admixtures have been developed as well as special methods of colloidal mixing. AWA may be a solution to the elimination of bleed. When the grout from anyone level of pipes has reached above the level of the second set, the grout injection points are moved up. Slotted inspection pipes are often employed to verify the level of grout. Electrical resistivity probes may similarly be used.

In selecting the aggregate to be placed, preference should be given to cubical particles as opposed to flat particles, since bleed water tends to be trapped under the flat particles. The major concern with grout-intruded aggregate concrete is to keep the aggregate clean. Silt, organic growth, and sand particles must be kept out of the placed aggregate. When aggregate is delivered by barge or vessel, fines tend to accumulate on the barge deck at the bottom of the pile due to abrasion and chipping. The lower layer of such rock should not be placed but should be wasted or used elsewhere. Grout tends to follow the path of least resistance. Fine particles increase the friction head and thus prevent the full flow of grout around all the particles.

Growth of algae can be minimized by covering the cofferdam to keep out sunlight and by addition of an inhibitor.

Grout-intruded aggregate is an excellent solution for concreting around embedments and instrumentation where these must be kept within very close tolerances. The fluidity of the grout, an essential quality, unfortunately also makes it flow out of any gaps in the forms and up through the exposed surface into the seawater. Hence, the pour should be as tightly enclosed as possible, while, of course, allowing the displaced water to escape.

Concrete made with sand only or with sand and fine aggregate (8 mm maximum) has been successfully pumped from a mixing barge down the legs of platforms to form bells at depths of 150 m and more for the offshore platforms of the Ekofisk Field in the North Sea and the offshore industrial terminal at Jubail, Saudi Arabia, in the Arabian Gulf. Pipe sizes of 50 to 75 m were selected to maintain a substantial friction head as the concrete was pumped "downhill." A vacuum release valve at the top of the pipe prevented a vacuum forming due to too rapid descent of the concrete.

However, the widespread practice of pumping the tremie concrete to fill drilled shafts, cylinder piles, and cofferdams is not sound. For such placements, with pipes larger than 75 mm, the loss of head due to friction is seriously diminished. The result is that the concrete is discharged in cyclic surges under essentially the fluid concrete head of a full pipe. Contrast this with gravity flow from a tremie pipe open to the atmosphere and hence automatically adjusting to near balance with the external water head, which allows the concrete to flow out slowly and smoothly.

Experience shows that such pumping of mass concrete produces an unacceptable amount of voids, mud inclusions, and laitance. Its use is generally not recommended. Conversely, pumping of grout is a sound and effective method.

The development of gravity platforms has led to the need for suitable mixes and methods for filling underneath a large, usually flat base with grout of special properties. Generally speaking, the desired properties are homogeneity and completeness of filling, low heat development, cohesiveness, low bleed, and long-term stability. The strength and modulus of elasticity required are usually very low; properties equivalent to the natural seafloor soil at a depth equal to the skirt penetration are all that are necessary. High-modulus grout is undesirable as it may form "hard points," concentrated loads on the underside of the base.

The quantities involved on an offshore platform have been up to 10,000 m$^3$, so the logistics of supply, mixing, and placement in the middle of the North Sea are obviously severe. Therefore, saltwater mixes have been selected, there being no embedded reinforcement for which to be concerned about corrosion. One mix developed used cement, retarder, and a finely ground filler of limestone. Other mixes replace 50 per cent or more of the cement with bentonite, or fly ash. Of particular interest is a stable foamed mix which uses only cement and seawater plus the foaming agent and a stabilizer. Limestone powder may also be included.

The materials are usually delivered to the platform by pump and then mixed and fed into a gravity-flow hopper located well down in the utility shaft. From there, the mix flows through pipes to the ejection nozzles. One clever scheme is to suspend a short length of hose under the base slab, held in a horizontal altitude by a chain. The hose then tends to ride on the surface of the soil, ensuring that the grout will be injected at the bottom of the space under the platform.

Proper venting must be provided to allow the trapped water to escape. Usually overflow ports are provided at a few meters height above the base to allow visual verification that filling is complete. Electrical resistivity gauges or nuclear methods in the standpipe or overflow pipe can serve a similar function.

Because of the importance of underbase grouting, a model test should be made, using the full thickness and flow length but, of course, reducing the width. Measurements should include temperature rise and bleed. Examination should be made for weak layers, large trapped inclusions, large bleed voids at the top, and laitance; 100 per cent fill is not required, as long as any voids are small, *e.g.*, less than a few percent of the area, and discontinuous.

Excessive pressure can cause piping out under the skirts or even raise one portion of the platform. It could lead to local overpressure damage to the base structure. Measurement of volume is a secondary guide to the completeness of filling and possible losses due to piping under the skirts. Therefore, pressures and volumes should be monitored carefully.

A special case arises when the structure is founded on a preplaced stone fill. The decision has to be made whether it is preferred to intrude grout into the stones or to have a grout that will not penetrate. Too fluid a grout may escape into the sea through passages in the rock. A suitable grout was developed for filling under the bases of the offshore caissons of the Hay Point Terminal, Queensland, Australia, using sand, cement, and a methocel admixture, which gave significant thixotropic properties. Tests indicated little tendency to segregate and little penetration of the rock. If the underside of the base is essentially flat, it may be desirable to provide inverted channels, which will ensure that the grout will flow under the entire base. The use of a grout with AWA or silica fume appears to be applicable.

Sandfill under the base is not strictly grout but is used for similar purposes. In the past, with subaqueous vehicular tunnels (tubes), sand has been placed in a semifluid state on one side and allowed to flow down, under, and part way up the other side. The scope of each operation has to be limited in order to prevent the fluid sand from raising the tube. For wide, flat-bottomed structures, Danish and Dutch engineers have developed sand-flow systems in which fluidized sand is pumped down under low head, to exit at distributed points under the base slab. As the sand spreads out in a circular pattern, the velocity drops and the sand is deposited. Any low spots become channels through which sand continues to flow until they are all filled. Thus the system leads to automatic complete filling. This system was employed to intrude sand fill under an exploratory drilling platform, the SSDC, in the Canadian Beaufort Sea.

A somewhat similar flow phenomenon is believed to be attainable with the thixotropic grout mixes, assuring relatively complete filling under the base.

Grout is extensively used to "cement" the annulus between pile leg and jacket sleeve. An annular gap of 50 to 100 mm is usually selected. The grout should flow from the bottom up. The mix is generally cement plus water. Fly ash may be used to replace part of the cement in order to reduce heat of hydration. Silica fume may be added to increase strength and reduce bleed. Admixtures may be used to provide water reduction, retardation, and expansion characteristics. AWA is an appropriate admixture. It is important that trial batches be made to ensure that the grout has the proper flow characteristics as well as strength. Flow rate should be kept low to avoid entrapment of voids. Grout should be overflowed to ensure that the initial mixture of cement and seawater is cleared. Pressures should be carefully controlled to prevent forcing the grout out from under the jacket sleeve; usually this exit is restricted by a grout retainer, but many times the grout retainer will have been damaged during pile driving. Therefore, a second entry grouting pipe is often provided, to permit the first grout to set and form a plug; then the main grouting is carried out through the upper entry port.

There are so many variations in the size, shape, conditions, and properties desired of underwater concreting that it is important

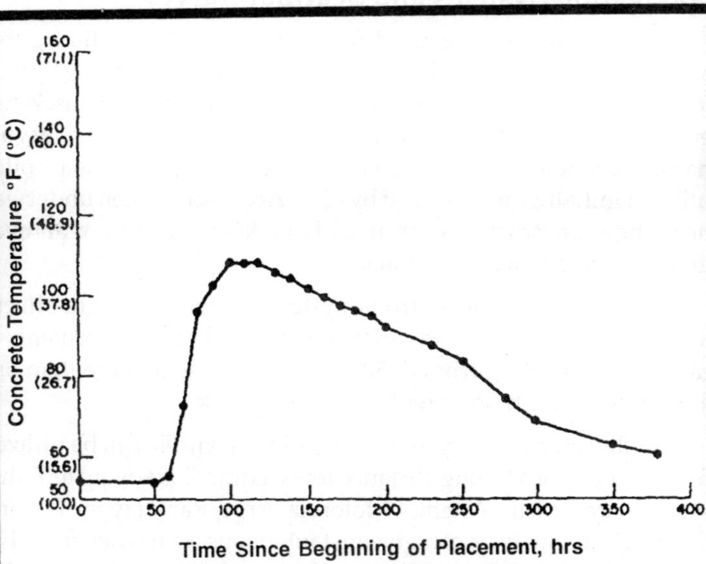


**Figure 10.1: Typical Curve of Heat Development in Mass Underwater Concrete Placement**

that recommendations such as those given earlier be treated as a guide only. Test mixes and trial runs are always advised to ensure that the best possible selection of mixes and methods has been made and that the personnel actually carrying out the work are cognizant of the precautions pertinent to that particular project. Both the mix design and placement procedure are all-important.

Most underwater concrete for marine operations is placed in a substantial mass. Therefore, it develops a high temperature which dissipates slowly. The external surfaces, sides and top, lose heat most rapidly, while the core still remains hot. When the temperature differential reaches 20°C over a thickness of 300 mm, the concrete will crack. Insulation of the sides by soil and of the top by a thermal mattress will reduce the temperature differential. Reinforcing steel should be provided, especially in large footings and those which are supported on piles, since the piles resist overall contraction of the newly hardened concrete. The steel area should be equal at yield to the tributary concrete width times depth of about 300 mm at the cracking strength of concrete.

## Navigation Systems and Seafloor Survey

Navigation systems used for control of position during tow and emplacement at the site include both Global Positioning System (GPS) satellite fixes and radio navigation positioning systems. When entering a site in which structures already exist and when leaving harbor, theodololite and electronic distance (range) systems are often utilized initially, then checked by GPS. Accuracies when underway near shore or structures can usually be kept to ± 1 m, with even greater accuracy when stationary.

Many long-range electronic systems suffer from night effects, losing accuracy. They also can be misinterpreted by increments of range steps, giving errors of 50 m. Thus the use of more than one system is desirable, in order to provide a check.

In the open sea, a system such as Decca Hi-Fix can be utilized for close control; for long-distance tows, Loran C is adequate. Other systems are Sydelis, Artemis, Motorola, Argo, Racal Hyper-Fix, and Omega. Satellite fixes can give instantaneous accuracies from 1 to 10 m depending on how many satellites are interrogated. The GPS, which became commercially available in 1986, now gives positioning accuracy to 1 m worldwide. Differential GPS, along with progressive declassification, continues to increase the accuracy. Short-range positioning systems include Motorola Mini-Ranger, Honeywell Micro-Automatic Station Keeping System, and Simrad.

Underwater acoustic transponders can be preplaced on the seafloor and used to control the final installation of a structure in the open sea. Usually six are preset to ensure that at least three will be working when needed. The use of these for final bathymetric surveys and for borings ensures that when the structure is finally installed, it will be at the correct relative position. They can also be utilized for the underwater assembly of elements and for guidance of an ROV.

Acoustic transponders have proved very satisfactory with steel jackets, but less is with massive concrete structures, which along with the many boats in the area, create excessive noise. A major acoustic transponder system has been developed by Ocean Instruments to enable placing of a jacket near existing pipelines. The acoustic positioning system is calibrated with satellite receivers to process Doppler information and thus provide a surface position fix every 1 to 2 hours.

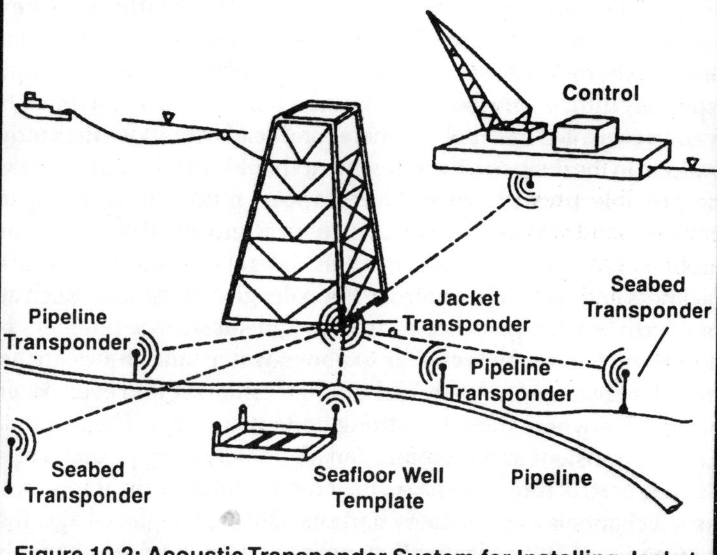


**Figure 10.2: Acoustic Transponder System for Installing Jacket Over Template Among Existing Submarine Pipelines**

Accuracy in the placement of offshore structures on the ocean floor has been steadily improving with the advances in equipment and with experience. In the early and mid-1970s, distances off target averaged 25 m, but by 1999 this tolerance had been reduced to less than 1 m.

Positioning of important structures should always be backed up by more than one system. During the installation of the Tarsiut Island, a survey tower was installed to control the screeding of the seats for the caissons, which was carried out with considerable accuracy. In order to screed the last base, the tower had to be removed and replaced. Due to a combination of human and electronic error, a gross error, occurred, resulting in the caissons being installed 20 m off from their intended locations, on an unscreeded bed. In hindsight, use of a secondary backup system of spar buoys, or even taut-moored buoys, would have prevented such gross error.

Seafloor surveys should be carried out in the vicinity of all marine and offshore structures, as well as along the route of submarine pipelines and cables. While the continental shelf is relatively flat and level, both the deep seabed and the coastal areas are subject to abrupt changes and anomalies. Similarly, many harbors are characterized by relatively flat bottom sediments but riverine bottoms are often highly irregular. Seafloor survey assessments should be made to disclose slumps scarps, irregular topography, rock outcrops, and the character of the seafloor material. Especially difficult are those sites where a thin layer of soft sediments overlies cemented material, or where coral heads rise above the sandy seafloor. In the deep sea, the assessment should additionally address the possible presence of mud volcanoes, mud lumps, collapse features, sand waves, slides, faults, diapers, erosional surfaces, gas bubbles, gas seeps, buried channels, lateral variations in strata thickness and subsea permafrost. For paleoglacial seafloors such as the North Sea the presence of surface and subsurface boulders is important. Sand waves similar to above-water sand dunes are an important feature for risers, harbors, and estuaries, and even occur far offshore where there are strong bottom currents. The fact that they are transient and therefore can alternately bury or uncover a pipeline or structure base makes their disclosure critically important. Buried channels exist in many harbors. During the glacial age, the sea was about 100 m deeper than now, so rivers cut deep channels out to the lowered sea. Windblown sand and volcanic ash were often deposited on the exposed surfaces. Subsequently, they have been flooded by the rise in sea level rise and the sharp topography filled by unconsolidated and weak sediments. Similar phenomena have created a series of caprock strata off coasts such as the North West Shelf of Australia. Of special concern to both engineers and contractors are soils containing greater than 15 to 20 per cent carbonates and soils containing mica.

Calcareous sands are composed of skeletons of minute organisms. Prior to fracture, these have relatively high friction angles and bearing strength; they behave like sand. Upon fracture, however, they behave as weak clay. Skin friction drops close to zero. This behavior enables piling to be driven with relatively low resistance. It also greatly reduces the pullout capacity. Direct bearing strength remains satisfactory. The presence of substantial quantities of mica not only reduces skin friction but also leads to instability of slopes and cuts. Finely divided mica is very difficult to ascertain, since minute particles do not show up visually. They can only be determined by physical-chemical testing.

Sands are very difficult to sample properly, and even the best samplers do not show the full density and consolidation, since the act of sampling and the reduction in hydrostatic pressure as they are raised to the surface invariably reduce their strength and apparent consolidation. Siltstone and mudstone are subject to water softening. What appears as soft rock may slack and disintegrate in the presence of water. Hence, borings, especially wash borings, may be reported as "silt and mud" but may actually be consolidated firm soils with the characteristics of soft rock. They often are resistant and abrasive in drilling or driving but are easily penetrated by a jet. Limestone strata may have been eroded to form solution cavities at times when the sea level and consequently the tributary levels of the rivers were up to 100 m lower. They may later have been partially filled with loose sand and silty sediments, then covered over with sands and even subsequent strata of cemented material. Finding these and defining their outline is a very difficult task, since borings may either miss them completely or conversely encounter a small but deep cavity, both giving an erroneous picture of actual conditions.

In many coastal structures, such as offshore terminals and outfall sewers, and in inland waters, shore-mounted lasers or even concentrated lights may be set up to provide a range which is directly visible to the barge superintendent and operator. This range, combined with electronic distance systems (EDS), may be used as the primary control, especially if frequent moving is involved. The advantage of these is that they enable the barge captain to judge rate of change as well as verify final position, and thus avoid excessive "hunt." Theodolites and electronic positioning were used to position the pylon pier caissons for the Great Belt Bridge in Denmark, with a final check made by repeated GPS observations. These caissons were set to an accuracy of less than 100 mm.

Bathymetric surveys are carried out by both death-finder sonic equipment and side-scan sonar. These must be corrected for roll, pitch, and heave and integrated with positioning systems. Such an integrated instrumentation system, called a "profiler," can give a plot of contours within a 200- to 400-m-diameter area. Such a system can be used for a wide variety of river, harbor, estuary, coastal, and offshore surveys. It can also develop images of large pipe, *e.g.*, outfall pipe segments and other objects on the seafloor.

Depth-finding sonar should be run at two frequencies, high and low, to detect the presence of soft, semifluid sediments overlying a firmer bottom. In a deep fjord in Norway, for example, use of the standard low-frequency sonar depth finder gave a depth 25 m greater than actual, since the acoustic waves penetrated the very soft soil without reflection. In the deep sea, sub-bottom profilers (tuned transmitters) can be used to determine both the bathymetry and the near-surface features.

In areas of strong relief, with steep or near-vertical bluffs and underwater canyon sides, the sonar echoes may come back from the side walls, indicating less depth than the true value. This is because of the conical spread of the beam. Narrow beams can be used to minimize this problem. An ROV with side-scan sonar was used very successfully to survey the bathymetry immediately behind Shasta Dam at a depth of over 100 m. Multibeam SWATH systems have been specifically developed in order to accurately depict the irregular seafloor in the deep sea.

Side-scan sonar can produce an excellent two-dimensional portrayal of the seafloor, along with any man-made objects such as pipelines, dropped objects, anchors, even anchor drag marks. Advanced acoustic imaging can now give a map of the seafloor with definition of less than 1 m.

Advanced photogrammetric techniques, using multiple photos, enable a small area of seafloor to be mapped to an accuracy of 25 mm in relief. Recent development by NASA of extremely sensitive film (ASA 2,000,000), combined with the use of strobe lights, has revolutionized optical seafloor search and survey.

Many new underwater acoustic systems and magnetometers are now available. Many of these can be fitted to an ROV and the data transmitted by telemetry back to the tending vessel. Others are used by divers. They enable the detection of buried cables and pipes,

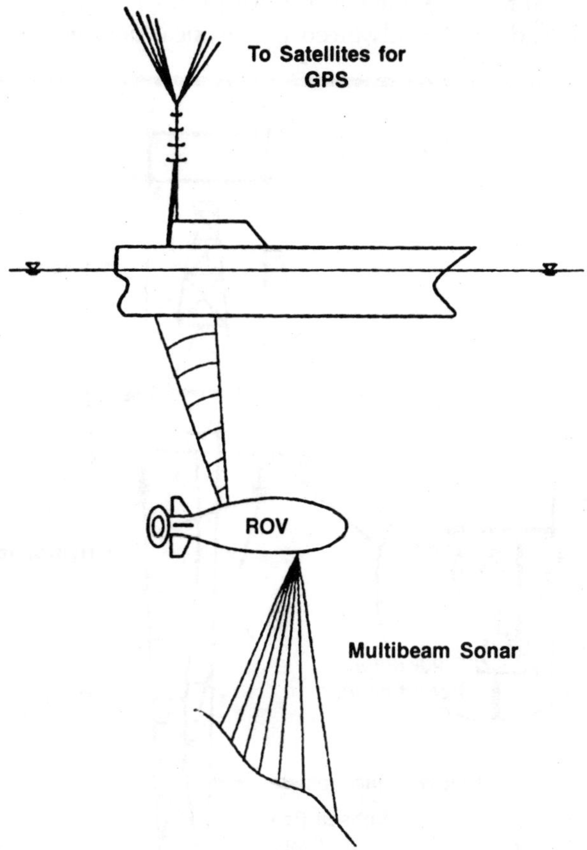


**Figure 10.3: Use of Captive ROV and Multibeam Sonar to Profile Sloping Seafloor**

leak detection, high-resolution acoustic imaging of the seafloor, range measurement for short-distance ranging, and guidance for entry of mating cones and piles.

Position-sensing devices and systems have been developed to enable a vessel to maintain station over a fixed position on the seafloor.

Sparker surveys can be run to determine the surface of subsurface hard layers and bedrock up to 100 m below the surface. Boomers can simultaneously be used for definition of surfaces to a depth of 100 m below the surface. Air gun, water gun, or sleeve exploder, and similar advanced geophysical devices can detect

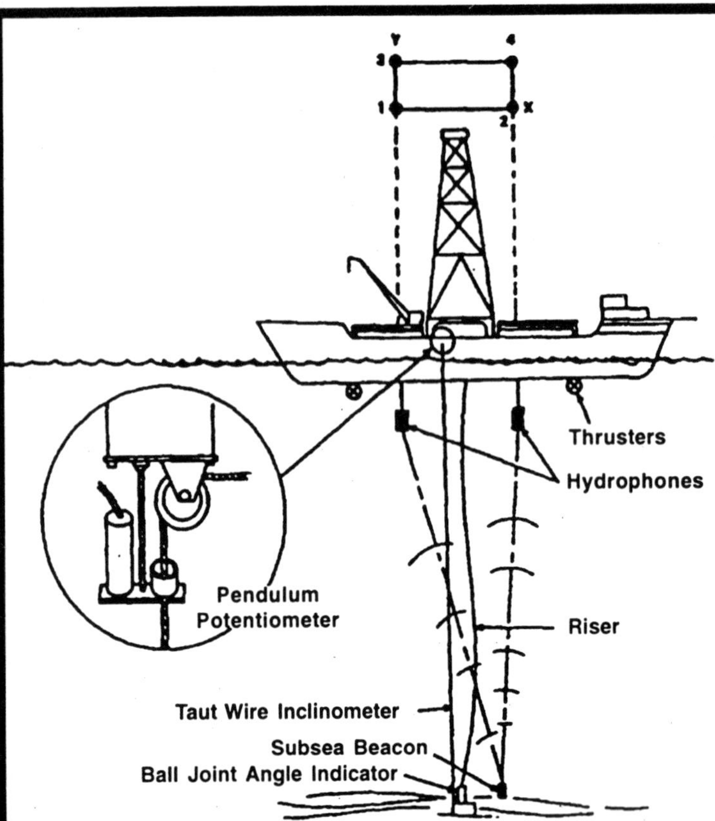


Figure 10.4: Position-sensing Systems for Maintaining Station

anomalous profiles at deeper penetrations (above 100 m.).

Another use for advanced surveying equipment is for detection of sea ice in the Arctic and sub-Arctic. Not only is mapping required but also indications of size and thickness of floes and the character of the ice, whether first-year or multiyear. These can be given by visual systems or by side-looking airborne radar (SLAR). The most advanced version of this is synthetic aperture radar (SAR), which permits five times better resolution than SLAR and can "see" through the cloud cover which so frequently obscures the ice.

Satellite coverage is now able to give all-weather, through-cloud coverage, through the use of SAR, identifying ice concentrations, presence of multiyear ice or ice floes, and presence of ice islands. SLAR and SAR flown by aircraft are available today, but coverage is limited in extent.

For underwater mating–for example, assembly of underwater structures, positioning of a jacket over a pre-installed template or of an articulated tower over a pre-piled base–a number of systems have been employed. Generally, the control barge, usually a large offshore derrick barge, is preplaced on a taut mooring. Side-scan sonar, electronic positioning, and acoustic transducers are used to verify its position relative to the underwater element. As the structure itself is lowered, short-range, narrow-beam sonars on the structure are used to interrogate acoustic transponders on the submerged element. A video camera with high-intensity lights, mounted on the legs of a jacket, may be used to verify position at close range.

ROVs may be used to verify conformance visually; their value is in preventing gross error due to some electronic reflections or misunderstanding on topside.

# Chapter 11

# The Seafloor of the Ocean and the Nature of Marine Soil

## Introduction

The seafloor of the ocean is highly complex due to its geological history and the action of the various elements, especially in the relatively shallow waters of the continental shelves. These shelves vary in extent depending on whether the margin is rising or slowly subsiding. Thus the east coast of the United States has a very wide continental shelf, whereas that on the Pacific coast of South America is very narrow. Beyond the continental shelves are the slopes, averaging 4° of slope down to the abyssal plain. Submarine canyons, which cut through the shelf and slope, may have side slopes as great as 30°. They usually terminate in a fan on the deep seafloor.

The ice ages have had a very dramatic influence on the shelf areas. When the Wisconsin ice age was at its peak about 20,000 years ago, immense quantities of water were withdrawn from the sea, lowering the sea level as much as 100 m. This meant that the shelves were exposed to this contour: beyond the shoreline of that date, seas were shallower than at present. Rivers discharging from land cut troughs into that contour and a little deeper, which is why so many entrances to sounds and other large bodies of inland water are now approximately 100 m deep. On these coastal shelves, land erosion processes took place. The rivers were steeper and velocities higher, and hence sedimentary deposits were coarser. As the oceans have risen, the velocities have been reduced, and finer sediments–sands and silts–have been deposited on the shelves opposite large rivers.

During this ice age, glaciers extended far out into what are now ocean areas, carving deep trenches such as the Norwegian Trench in the North Sea, Cook Inlet, Alaska, and the Strait of San Juan de Fuca between Washington and Vancouver Island. With the warming period in which the world now finds itself, the sea levels have been rising, slowly but inexorably flooding coastal areas, changing drainage patterns, creating shoreline features.

Glaciers retreated, leaving morainal deposits. Shallow freshwater lakes were eventually inundated, trapping their land sediments. Rivers dropped their sediments sooner, creating deltas through which new channels were cut. Volcanic ash fell through the shallow waters, as did wind-blown (aeolian) sand. The loosely deposited sediments have been subject to mud slides, triggered by storm waves or earthquakes. Turbidity currents have removed millions of cubic meters of coastal sediments, to flow downward to form a fan on the abyssal plain, creating great submarine canyons during their flow.

Most structures in the ocean extend over substantial areas. There may be significant variations in soil properties over this extent. Because of the cost and time required, it may not be possible to obtain a sufficient number of borings to show the true situation with its variations. There is a tendency to place undue emphasis on the few borings that may be available. Geophysical methods such as "sparker surveys" and a study of the site geology may help to alert the constructor to the range of soil properties that may be encountered.

The continental shelf is typically smooth and featureless, with a gentle slope. In contrast, the deep sea can be rugged and highly variable. The geologic process include landsliding, active faulting, and seabed erosion. The topography of the deep seabed of the Gulf of Mexico is rough and irregular due to past and ongoing uplift of salt deposits. Vertical uplifts may be 2 to 4 m over a 100-year period. Rocky seafloor caprock deposits have formed and gas hydrates have been trapped. Hydrate mounds 300 to 500 m across and 40 m above seafloor have been encountered. Biological communities form on the salt uplifts and at seeping hydrocarbons. The salt uplifts may be several kilometers across and up to 200 m above the surrounding seafloor. Fault scarps can be as steep as 45°.

The properties of deep-water clays may be altered biologically and become much more sensitive than conventional marine clays. Brine lakes may have significantly greater fluid density. Chemosynthetic communities such as tube worms occur at seafloor vents and at hydrate mounds and faults from which hydrocarbons are seeping.

The pipeline crossing of the Strait of Gibraltar at a depth of 300 m encountered many abrupt scarps 10 to 20 m in height.

Estuarine and many harbor bottoms typically consist of very fine sediments such as muds, clays, and sands. The upper sediments are recent deposits, hence loose and weak. They may be of variable depth, filling river and creek canyons which were formed during the post-glacial age when the sea level was lower. These loose fine sands and silts may be subject to flow slides during construction and dredging, apparently due to excess pore pressure building up on the uphill side, with failure initiated by a local vertical face formed during dredging, triggering instability. Flow slides may occur on slopes as flat as 1 on 6 or even flatter.

River deposits vary from recent coarse sands to layers of gravel and cobbles, deposited during floods. Lake deposits are typically fine sands and clays.

## Sands

Many river beds comprise primarily sands. Sands are extremely hard to sample during geotechnical investigations.

In many offshore areas there are very extensive accumulations of sands, in some cases due to longshore sediment transport of sand discharged from rivers in other cases due to ancient sand dunes.

Scour from wave action can occur when the water depth is less than half the significant wavelength. Whenever the depth is less than one fourth the wavelength, substantial scour in sands is likely. Bottom currents, whether wave induced or from other causes can move sand, especially if it is periodically raised by the pore pressure gradients induced by waves. The resultant scour holes tend eventually to stabilize at a condition where the rate of sand infill matches the erosion rate.

River and tidal currents adjacent to structures produce scour, especially around the corners and underneath a structure. Blocking of the river by equipment and structures such as cofferdams amplifies the velocity and scour potential.

The excavation of underwater trenches in sands is typically rather difficult and complex due to the varying densities of the sands. However, stable side slopes can be achieved in the absence of severe currents and wave action at shallow depths. $N_{SPT}$ values provide a guide. $N_{SPT}$ values obtained underwater must first be multiplied by 1.12 to account for the effect of submergence. In cohesionless sands, $N_{SPT}$ values must also be corrected for depth.

**Table 11.1: Correction Multiplier to Apply to Measured SPT Values (as adjusted for submergence) to Account for Overburden Pressure at Various Depths**

| *Depth Below Seafloor (ft)* | *Correction Multiplier to Give Value Under Standard Confinement Pressure* |
|---|---|
| 2. | 2.3 |
| 5 | 2.0 |
| 10 | 1.8 |
| 15 | 1.5 |
| 20 | 1.3 |

**Table 11.2: Correlation Between SPT and Stable Dredged Slopes in Cohesionless Materials (*e.g.*, Sand)**

| | *Very Loose to Loose* | *Loose to Medium* | *Medium to Dense* | *Dense to Very Dense* |
|---|---|---|---|---|
| Corrected $N_{SPT}$ | 0–4 | 4–10 | 10–30 | 30–50+ |
| Relative density | 0.05 | 0.35 | 0.65 | 0.85–1.0 |
| Moist unit weight (lb/ft) | 70–100 | 90–120 | 110–130 | 120–140 |
| Stable slope | 4 : 1 | 2.25 : 1 | 1.75 : 1 | 1.5 : 1 |

The construction of a structure on sands modifies their behaviour. Wave energy acting on the structure is transmitted to the foundation, increasing the pore pressure. Due to its cyclic nature, the pore pressure may be progressively built up until local liquefaction occurs under the edge. This eventually leads to a loss of material under the edge and the tendency for the structure to rock, aggravating the problem. This is why concrete caissons used as coastal seawalls usually fail outward under wave attack: they have lost the sand under their toe.

Coastal sands move laterally under the action of the prevailing current. Structures which interfere with that movement cause the sand to build up on the "upstream" side, and erode on the downstream side.

Under the action of strong currents such as those found in the southern North Sea and at the mouth of major rivers of South America and Southeast Asia, the sand bed may be formed into waves, that is, sand dunes. These dunes move just as their landbased counterparts do, eroding from the back, redepositing on the front. Typical maxima heights are 3 to 10 m. Thus in planning installations in such areas, it may be necessary to dredge the dunes down to or below the level of the troughs; otherwise the structure or pipeline may end up exposed above the seafloor.

Sand dunes also form in the beds of rivers and estuaries, moving downstream with the prevailing or dominant current. Sand waves up to 10 m in height move downstream in the bed of the Jamuna River in Bangladesh at a rate of a few kilometers a day.

Sand deposits in the North Sea and off Newfoundland have been subjected to continuous pounding by the storm waves above. *Pounding* is perhaps an inaccurate term. What does happen is that the internal pore pressure in the upper layers of the sand is alternatively raised, then drained, only to be raised again. Pore pressure variations of 3.5 T/m$^2$ (35 kPa; 5 psi) have been measured. After millions of cycles, the sand becomes extremely dense, often with consolidation higher than can be reconstituted in the laboratory. Friction angles in excess of 40° may be found.

When sampled by conventional techniques, the sands are automatically disturbed; hence laboratory tests will often under-report their density and strength.

## Underwater Sand Deposits

The offshore areas of principal interest to the petroleum industry are great sedimentary basins. Although many are ancient deposits and relatively stable, others are still active deltaic areas. The great freshwater rivers–the Mississippi, and the rivers of Asia, South America, and Africa–are transporting huge quantities of silts and clays in colloidal suspension. Contact with the saltwater causes flocculation, and the highly dispersed soil particles settle to the seafloor. Periodically, huge blocks of these recent sediments slump off and flow seaward. This process is prevalent even far out to sea, on the sides of submarine canyons such as the Baltimore Canyon of the U.S. East Coast and the outer edge of the continental shelves. Evidence of widespread slumping exists near the break of the continental shelf of the Alaskan Beaufort Sea. In the Gulf of Mexico, slumping occurs as mud slides.

Underwater sand deposits may also be very loosely consolidated. If the internal pore pressure is raised by any of several mechanisms, the sand turns into a heavy liquid. These underwater flows and turbidity currents have occurred in both sands and clayey silts. Some of these occur frequently, and are the mechanism by which shore sands are fed down submarine canyons, to be deposited in the fan at the bottom of the continental slope. During this downward flow, they erode the canyon itself. Others are more infrequent, being triggered by an intense storm or by earthquake. In clay areas, these slides are aggravated by entrapped methane gas in the silty clays. These failures often occur on very flat slopes, which superficially would appear to be stable.

While the natural occurrence of these would appear to be, like earthquakes, important to the designer but not to the constructor, there is one important difference. They may be and have been triggered by the construction operation itself. Pile driving, dredging, dynamic compaction, and underwater rock dumping have triggered flows in the Norwegian fjords which have involved millions of cubic meters of soil and have extended out into the offshore areas, eventually pouring into the Norwegian Trench.

Temporary cofferdams and piers constructed off the coast of California have transmitted wave energy into the sands, triggering slides 2 km wide, again involving millions of cubic meters of sand which were transported down into the Monterey submarine canyon. In one case an existing pier "disappeared" in another the sheet pile cofferdam was lost. Turbidity currents off the U.S. East Coast have been known to sever submarine cables many kilometers offshore. These submarine slides and turbidity currents are typically limited as to the depth of soil involved to about 20 m.

As petroleum exploration and hence offshore construction extends out to the edge of the continental shelf and beyond to the slope and eventually the fans themselves, these submarine flows will become of increasing concern to the construction operations as well as the design.

Bathymetry at and adjacent to the construction site is extremely important, as it affects initial setdown of jackets and caissons as well as the landing of more flexible structures such as pipelines. Adequate surveying methods and locating systems must be employed to ensure that the bathymetry, geotechnical, borings, and actual installation are all controlled to the same positioning relative to each other. Because of tolerances and systematic errors (*e.g.*, night effect) on many electronic survey systems, a method which marks the true position of the site–for example, GPS, acoustic transponders, or, in shallow water, articulated spar buoys–will usually be found desirable.

The site survey should also identify carefully the relative position of any foreign or artificial objects–pipelines, abandoned anchors, dropped casing, and the like–which often litter the seafloor in the vicinity of an offshore site due to the prior conduct of exploratory drilling operations. Side-scan sonar and ROV video are usually the most effective means.

Scour and erosion are addressed in specific sections where their occurrence is most likely. The potential for scour exists at depths up to 100 m and even greater, where wave action may build up internal pore pressures and where eddy currents may have vertical as well as horizontal components. Recent studies indicate that scour potential may even exist in certain areas of the deep ocean where periodic eddy currents are superimposed upon steady-state unidirectional currents. Since scour may occur during or immediately after the installation of a structure, it is essential that monitoring is carried out and adequate protection placed as soon after landing as possible. In some cases it may be necessary or desirable to place scour protection prior to the installation. When a large caisson is being seated on the seafloor, for example, the water trapped underneath must escape, thus creating a scouring action. At this same time of installation, when there is a relatively small gap under the structure, current or wave action may induce a high flow rate under the structure, causing scour.

A similar phenomenon is that of piping. Piping is the formation of a channel or tunnel under a structure due to pressure gradients which erode the soil locally. Such piping not only weakens the foundation but also may prevent subsequent construction operations which require the maintenance of an overpressure (*e.g.*, for removal of the structure) or underpressure (*e.g.* for aiding penetration of skirts).

Certain soils may degrade and soften when exposed to the changed conditions they experience during construction, especially if they have been previously blanketed by impervious material or were loosely cemented.

In summary, it may be fairly said that seafloor geotechnics represents the area of greatest concern to and difficulty for the constructor. Problems of instability, inability to penetrate, and slope failure continue to plague marine construction activities. Overconsolidated silts in the Arctic and calcareous sands in the subtropics are perhaps the most demanding problems facing the constructor in today's offshore operations while loose silty and micaceous sands pose great difficulties for river and harbor construction. Thus a closer relationship between the geotechnical and construction engineers should lead to more effective and economical offshore construction.

## Underwater Activities

Recent years have seen a revolutionary growth in society's concerns about the impact of activities, especially construction, on the ecology as well as on the health and quality of life of humans. Marine construction activities take place in an extremely sensitive environment, since water readily conveys local discharges and effects to the wider area encompassing, in extreme cases, an entire estuary, bay, or even a sound, *e.g.*, the mammoth oil spill in Prince William Sound, Alaska. Public concern has focused on marine activities, resulting in a host of regulations intended to eliminate or mitigate damages to the ecology and to minimize the danger and disturbance to human communities.

It has become of great importance to incorporate these rules and precautions in the planning stage rather than, as in the past, attempt to correct or mitigate the negative effects during actual construction. Constructors need to consider these rules as an inherent requirement to their work, similar to the specifications, but with the added force of law.

The concept of ecology is that of an all-inclusive living system, ranging from microorganisms to whales and including humankind. Any disruption of this system may have a chaotic effect, causing extensive negative effects throughout the system.

The construction contractor is generally not involved with drilling for oil but may very well be conducting operations in the vicinity of live oil lines. Thus the constructor may well have the potential for causing an oil spill. The constructor's operations themselves involve the use of fuel oils (diesel, gasoline, etc.) and lubricants. Leaking equipment, errors in transfer of fuel, failure to close and seal valves may all produce the "sheen of oil on the surface" which is prohibited by regulations in many coastal waters and is unfortunately highly visible from the air. The amount of oil that is tolerable is, of course, the subject of highly emotional debates. However, there is no question but that this is a matter to which construction contractors must give attention and that they must take active steps to prevent oil spills.

The most harmful immediate effects of oil spills are the contamination of the feathers of seabirds. Oil may travel long distances, eventually ending up on a beach where it has serious aesthetic as well as some harmful biological effects. Fortunately, these latter do not seem to persist for long on active shoreline beaches. Oil in estuaries, marshes, and the like appears more harmful. Another serious effect is the contamination of the beaches and shoreline rocks where lesser marine organisms such as mussels, sea anemones, and algae thrive. Since these are an essential part of the food chain, oil deposits are harmful. However, the use of steam cleaning and detergents may be even more harmful. Gasoline and diesel oil are more toxic than crude oil. Oil, being an organic substance, biodegrades in the open water, due to a combination of bacterial activity, oxygenation, and sunlight.

Many laws and regulations restrict operations that can endanger the breeding sites of marine mammals and the nesting sites of birds. Operations which interfere with migratory routes of fish will be restricted. Seafloor disturbance of colonies of shrimp or mussels is a concern. Many of the above are seasonal, others are specific as to water depth and location.

Use of explosives underwater is severely restricted because of fish kill. Surface blasting (bulldozing) is the most damaging. Shaped charges may produce less kill if they are mounted in a frame or are otherwise secured so that they do not become dislodged and turned over. The best method, with minimal fish kill, is by drilling and blasting, with packing to keep the explosion contained below the seafloor.

Noise affects the nearby breeding of marine mammals. Turbidity can ruin oyster and mussel beds. Oil is very damaging to birds and to a lesser extent to fish. The subsequent cleanup by steam cleaning and detergents may be even more damaging.

Several endangered species of birds, such as the least tern, nest on coastal shorelines and in wetlands adjacent to rivers and harbors. Construction during the breeding and nesting season may be severely restricted or prohibited.

In many areas, fish migration is very close to shore, in relatively shallow and protected waters. An example is along the west and north coasts of Alaska. In other areas, fish and migrating mammals often use a relatively narrow channel. In rivers there are seasonal restrictions on work in shallow water, through which the salmon and other androgynous fish migrate. In some shallow-water areas, algae and sea grass grow which is the food of endangered species of fish.

For such limited and constricted areas, even relatively minor disturbances can have important consequences. For example, an offshore jetty would obviously interfere, but so also might an overhead bridge, since many fish, such as salmon, reportedly are reluctant to swim under a shadow.

Most countries today require the filing of an environmental impact statement or report prior to the undertaking of a major marine project. This will usually have been filled by the client. Included will be sections dealing with the impacts during the construction phase and the marine and onshore impacts of marine operations. It is important for the offshore construction contractor to become familiar with these documents and the constraints, restrictions, and mitigating procedures set forth in them. Compliance is not only legally required but, as a practical matter, is essential in order to assure that the construction operations may proceed without interruption or delay. Lack of strict compliance may involve the constructor in legal disputes, even criminal charges, and in today's social and political environment, may stir up public opposition and interfere in the operations.

High-speed outboard motors, helicopters, low-flying aircraft, discharge of dredged gravel through sub-marine or floating pipelines, pile driving, drilling, sparker and seismic surveys, and even echo sounding are examples of construction operations which create noise in the water column.

Noise appears both to attract and repel sea animals. Low-frequency noises travel farther in water. Concern has been expressed that wideband noise spectra may interfere with the navigation used by the bowhead whale. There is concern among the Inuit hunters that the noise may drive the whales and seals farther offshore, to the edge of the polar pack, where hunting is more difficult and dangerous. Several experts believe that the distance over which construction and drilling noise will have a significant effect, is of the order of 1000 m.

Noise may be isolated by an air gap, such as that created by intense bubbling of air around the resonator in contact with the water. Ma, Veradan, and Veradan in OTC Paper 4506 (Offshore Technology Conference Preprints, 1985) have shown that gas or air bubbles in water and sediments can attenuate long-range, low-frequency underwater sound propagation very effectively.

Many larger animals (caribou, geese, ducks, etc.) appear to become accustomed to the noise of helicopters if they are not too close, although it is generally believed that loud noise such as that from low-flying aircraft is injurious to breeding birds. Airborne noise may disturb calving of sea mammals on adjacent shoals. It can also be a significant nuisance to inhabitants of nearby shores, since noise, especially low-frequency noise such as that from pile hammers travels long distances (2000 m and more) over water. Typical restrictions on noise at an affected location are 65 decibels for 5 min every hour: from 6:00 a.m. until 9 p.m.; for the remaining period, 55 decibels is the maximum. Near waterfront hotels or residences, pile driving may be prohibited at night. Noise from gravel traveling through the disposal pipes of hydraulic dredging appears to be especially detrimental to marine mammals, interfering with their communication. Noise is, of course, a serious concern for workers at the site. Earplugs are required for workers in the vicinity of pile hammers and diesel engines. A number of means have been tried to minimize the noise of pile hammers. Wood or plastic cushions between the pile hammer striking block and the pile head are partially effective. Curtains and boxes have been tried; they are not very effective. Unfortunately, sound absorption requires mass.

Similar limits may be placed on bright lights near shore prefabrication sites. For example, bright lights were a major issue during the construction of the concrete offshore platforms in Stavanger, Norway, resulting in the construction of shields to protect nearby residences.

Highway traffic, as well as rail traffic, may be impacted by the new construction. This may result in detours, limited hours of work, and establishment of sequences for the several phases.

Navigation must be allowed to continue with minimum interruption and with minimal impact on safety. This usually constrains the scheduling of construction of overwater bridge piers and requires the construction of fenders, buoys, and navigation lighting. Notices to Mariners are issued by the regulatory bodies such as the U.S. Coast Guard to warn shipping.

In rivers and estuaries, the restriction of the river by cofferdams or other structures and by moored construction equipment will cause a small backwater rise in water level above and, more importantly, an increase in river velocity in the remaining cross section of the water flow. This may adversely effect navigation, especially upstream traffic. For example, tug-barge combinations may have a maximum still water speed of 5 knots (2.5 m/s). If we assume that the normal river flows 1.5 m/s, and the construction raises the local velocity by 1.0 m/s, then a helper tug will be necessary.

Downstream barge operations are also affected, although less severely. If a minimum speed through the water of 1.5 m/s is required to maintain control, then the increase means that the speed relative to the obstruction is now 3.0 m/s or 6 knots, making it difficult to thread the narrowed channel accurately, especially if upstream traffic is passing.

When working in the vicinity of airports, the height of booms and towers will be strictly regulated. Care has to be taken that in lofting (pitching) of piles, they do not extend into the critical air space, even momentarily, without specific permission from the air traffic controller in each case.

During construction of the Overwater Runway Extension of the La Guardia Airport in New York, as well as more recently during the construction of the Øresund Tunnel in Denmark, a boundary cone was prescribed in three-dimensional air space. The airports changed their normal landing patterns to use other runways whenever practicable, but takeoffs were directly over the construction operations.

Many offshore construction operations must be carried out in the vicinity of existing structures and facilities. For example, it is increasingly becoming the practice for the oil company to have wells drilled prior to the installation of the jacket and platform. Subsea satellite wells may similarly have been completed ahead of platform construction. Flow lines and pipelines may be in the vicinity. It is, of course, essential that these not be damaged by the construction contractor through carelessness such as allowing an anchor line to be wrapped around a subsea well completion or an anchor to be dropped onto an existing pipeline. These have occurred, with serious financial cost for repairs. There is always the possibility that oil will be released to the sea.

Particular care has to be taken when in the vicinity of seafloor well completions.

Pipelines and moorings laid in an active bottomfishing area have to cope with trawl boards and nets. Although these can damage the line, most often it is the fishing gear which is lost (or claimed to be lost), with resultant claims for reimbursement.

In the vicinity of saltwater intakes for onshore facilities such as LNG plants and power plants, sediments, especially sand, may be a hazard. Sand particles, for example, swept into suspension in an intake may clog spray nozzles in the plant. Operations therefore will have to be planned to minimize stirring up of the seafloor sediments. In extreme cases it may be necessary to install barriers on the seafloor (*e.g.*, steel frames with filter fabric curtains) to prevent sand movement just above the seafloor.

Considerations of existing installations require that very careful surveys be made prior to the start of operations and that their relative position be tied in to visible structures or acoustic transponders, so that they may be a guide to subsequent operations. Side-scan sonar or more sophisticated profiling systems are the usual means for location of underwater structures. On occasion they may need to be supplemented by underwater visual or video means using a submersible or ROV or by diver surveys.

River and harbor construction is commonly required to be carried out in close proximity to existing structures and may even connect with them. Usually they will be in operation and only limited incursion and shutdown will be permitted. New structures may be required to connect with existing structures.

It is important that the constructor of the new project have both as-built drawings and accurate survey information on the existing structure, since many older structures were not built to exact tolerances and field changes may have been made. Careful records should be kept of all anchor locations and the survey plots used in setting them to provide verification of the contractor's work and to protect the contractor from claims for damage that may have been caused by others also working in the vicinity.

The client or others may be carrying on other operations in the vicinity of the new construction: tanker loading, offshore supply, drilling, and other construction. One common source of problems is anchor line interference. In such cases, carefully planned schedules and layouts should be agreed upon by all parties and then adhered to. If it becomes necessary to adjust them, all parties should be notified promptly, as more than one may have planned operations in the same "weather window" and space.

When working in the vicinity of operating facilities, especially offshore terminals, there may be very strict limitations on operations (*e.g.*, welding) involving the possibility of fire or explosion. These may be conditioned upon the direction of the wind. Similarly, work near an operating flare may be dependent on wind direction in order to avoid excessive radiant heat. At loading and unloading terminals, the constructor may be required to shut down all welding and burning or even all operations while the transfer of petroleum products is taking place.

Dredging and excavation plans must ensure against the undermining of an existing embankment, levee, or seawall. Unequal removal or deposition of material may cause lateral displacement or settlement of an existing facility. When conducting blasting operations near or adjacent to an existing structure, the shock may be attenuated by air bubbling. Timber mats may be placed to protect valves. Blasting mats can be deployed to prevent airborne fragments.

## Boulders on the Seafloor Surface

*Boulders* on and near the seafloor are typically found in sub-Arctic areas where they may have been deposited by ice rafting. Another widespread process is that of erosion, occurring when the sea was shallower than at present. The weaker deposits eroded away, dropping the boulders down and concentrating them. A third process occurs in granitic soils such as those of the east coast of Brazil and west coast of Africa, as well as in Hong Kong. As these rocks have weathered into residual soils, resistant cores have remained firm, thus becoming "boulders" formed in place. Borings made in such residual soils usually miss most of the boulders. When they do encounter one, they often erroneously report it as "bedrock." The practice in Hong Kong, after much distress due to excessive settlement, is to core 5 m below the top of "bedrock" in order to confirm validity.

Boulders also exist in clay deposits. Some of these arose as morainal deposits from glaciers, discharging their bed load into shallow water muds, which have since been overconsolidated by subsequent advances of the glaciers. These are the boulder clays of the North Sea. *Glacial till* is a term used to describe these unstratified conglomerate deposits of clay, gravel, cobbles, and boulders found in many Arctic and sub-Arctic regions. The term is very nonspecific; some glacial tills have little binder and may be largely composed of gravel, and cobbles, whereas others may contain large boulders. Perhaps the most difficult are the well-graded tills, with all the interstices filled with silts and clays, so that there is a very low percentage of voids. These deposits are usually heavily overconsolidated, resulting in a high unit weight and a structure superficially resembling that of weak concrete. Thus unit weights have reached 2400 kg/m$^3$.

Geotechnical explorations, unless very carefully planned and carried out, often find only the finer sediments, and the samples will have been highly disturbed, resulting in the indication of much weaker material than that actually encountered. Wash borings bring back mud and sand. These tills are hard to drill; the material is very abrasive and hard, yet the bond is weak. However, high-pressure jets have proved effective in penetrating these tills. If relief can be provided in the form of holes or exposed faces so that the overconsolidation pressure may be released, then more normal construction activities such as dredging and pile driving may be carried out effectively.

Individual boulders and cobbles have not proved to be as difficult a construction problem as originally feared. A heavy-walled steel pile will usually displace them sideways through sedimentary soils. The same will occur when a large caisson with heavy steel or concrete skirts lands on a boulder. Clusters of boulders are a much more difficult problem. Where suspected, efforts should be made to locate them and remove them or relocate the structure accordingly.

Large boulders underneath the base of a caisson, for example, could exert a high concentrated local force on that base. Large seafloor boulders have been successfully removed from platform sites in the North Sea, by using trawler techniques, dragging the boulders clear. Another means used has been to place shaped charges to break them into smaller pieces. These boulders do not show up well on side-scan sonar or acoustic imaging. Work submarines (submersibles) taking video pictures, using special lighting, have been the most effective in determining the presence and size of seafloor boulders. Below-surface boulders can be sometimes located by sparker survey and, in shallower water, by jet probes.

Some seafloor areas which have been subjected to high currents or wave action are "paved" with cobbles, closely packed, with or without sand in the interstices. These cobbled areas are difficult to excavate because most conventional equipment has difficulty getting a "bite" into the material. Once a trench or hole is begun, the slopes may become very loose and unstable, taking a rather flat angle of repose, for example, 2 : 1 or even flatter, depending on the current. Their rounded surfaces give a low angle of friction. If a drilled shaft is to be constructed, it may be necessary to grout below the tip of the casing in order to stabilize the cobbles so they can be cut. Petrcussion drilling (churn drill or down-the-hole drill) may prove useful.

In sub-Arctic areas, deep deposits of gravels are found. These have been eroded by glaciers and rivers from the mountains and discharged into the sea to have a minimum of stratification and fines. Although composed of sound material, they often develop very low frictional resistance, due to their rounded surfaces and high void ratio. Thus they are quite unstable and dredged slopes will be very flat, for example, 3 : 1.

Piling driven into such material has typically failed to develop the desired skin friction. End bearing may also be less than expected, due to the large void ratio. It has proved necessary in many cases to provide enlarged tips or deeper penetration of piles in order to obtain adequate bearing. Such gravel deposits are inherently difficult to sample by conventional means, since any samples are more or less completely disturbed, and it is difficult to determine the degree of packing (consolidation).

The end weathering process of many rocks results in the formation of clays, which constitute the large bulk of many deltas. These clays consolidate under the overburden of later deposits. These clays are highly impermeable and cohesive. They are often anisotropic, with greater horizontal permeability than vertical. Often there are thin lenses or strata of silts and sands embedded in the predominant clay body. Clays usually contain organic material. The behaviour of clays is determined by their particle shape, mineralogical composition, and water content. Thin flat plates similar to montmorillonite possess dynamic lubricating qualities. Other types of clays may be sticky, "gumbo," plastic, or firm.

*Mud* is a term used to denote very soft, highly plastic, recently deposited clays. Typically, marine clays show shear strengths ranging from 35 kPa (700 psf) down to 14 kPa (300 psf), although some surficial muds may have only 2 kPa (50 psf). These qualities of clay and mud present a number of problems to the constructor. Among them, the following deserve special attention.

Clays tend initially to stand at relatively steep slopes when the excavation depth is limited. The buoyant weight of submerged clay is much less than the air weight; hence the driving force leading

**Table 11.3: Correlation Between SPT and Stable Dredged Slope in Cohesive Soils (Clays)**

| *Soil Consistency* | *Unit* | *Very Soft to Soft* | *Soft to Medium Stiff* | *Medium Stiff to Stiff* | *Stiff to Very Stiff* | *Hard* |
|---|---|---|---|---|---|---|
| $N_{SPT}$ | bpf | 0-2 | 2-4 | 4-8- | 8-16 | 16-32 |
| Typical depth[a] | ft | 0.1-10 | 15-25 | 25-40 | 48-80 | 80-100 |
| Shear strength (from $N_{SPT}$; or from unconsolidated, undrained laboratory tests or from field vane shear tests) | ksf | 0.25 (250 psf) | 0.5 (500 psf) | 1.0 (1000 psf) | 2.0 (2000 psf) | 4.0 (4000 psf) |
| | kPa | 12 | 25 | 50 | 100 | 250 |
| Stable slope[b] | | Requires special consideration | 4 : 1 | 1½ : 1 | 1 : 1 | ¾ : 1 |
| Need to consider surcharge? | | Yes | Yes | Possibly | Normally not required | |

[a] The depth of normally consolidated clay associated with the shear strengths shown.

[b] This is the ratio of horizontal distance to vertical height.

toward failure is much reduced compared to the same excavation in the same clay above water. With time, however, the clays strain (creep) and lose strength, failing in a typical curved shear plane.

Clays typically are penetrated rather readily by a pile under the dynamic blows of an impact hammer. Their short-term cohesion against the side of a pile is low. However, with a short period of rest, the soil will bind to the pile with its full cohesion, a process called "setup" or "set." Thus in driving piling in clays, when a stop is made to splice on another section, the blow count will jump up considerably when driving is recommenced. On occasion, it will not be possible to get the pile moving again. A portion of this increased resistance usually remains as permanent resistance.

Clays may present problems in dredging, due to their cohesive nature. In hydraulic dredging, clay balls may form. Flow will not be uniform. In bucket dredging, the clay may stick to the bucket and not discharge readily.

Once clay is in suspension, it becomes colloidal in behavior, and the discharge will be highly turbid. It requires a considerable length of time in relatively still water for clay particles to drop out of suspension. Where permitted, chemical flocculants, even seawater, will cause more rapid flocculation and dropping out of suspension.

Clays are normally quite resistant to scour. However, where strong bottom currents exist or where layered soils permit cyclic pore pressures to build up under overlying impervious strata, scour must be considered. Scour pockets in clays do not continuously refill as they do in sand. Weak muds in harbors and estuaries scour over a period of time, due to velocity increase around a structure. The scour of clay banks on the outer bends of rivers or where currents are high due to constrictions often proceeds by a process of undermining and collapse.

Now let us study about silts. Silts constitute one of the least-known types of soils, lying as they do in the size range between sands and clays and exhibiting properties different from both. Silts are typical of Arctic and sub-Arctic regions, although they also exist in temperate climates. They have been encountered in the Beaufort Sea, in Cook Inlet, Alaska, in the St. Lawrence Seaway, and off the coast of California, this latter case being a weak siltstone.

One unique property is their overconsolidation. The overconsolidation of clays, for example, is usually due to their having been subjected to intense loads from overburden or ice (glaciers) which have subsequently been removed by erosional processes or melting. Silts, on the other hand, have frequently been found to be overconsolidated even when there is no prior geologic history of burial. Various hypotheses to account for this include freeze-thaw cycling in shallow water, wave action, and electrostatic attraction. Regardless of cause, these overconsolidated silts are extremely dense and resistant to penetration, pile driving, and dredging.

Sampling and even *in situ* vane shear tests almost always disturb these silts. In many conventional borings, the silt will be reported as "mud." These silts are typically very abrasive to drills. Yet they break up readily under the action of high-pressure water jets. Paradoxically, some of these silts remain in suspension for long periods, but when they do settle out, they become very dense once again. To the contractor they pose problems similar to a very soft rock which degrades when exposed to water.

## Chapter 12

# The Structural Components of the Offshore Platform

### Introduction

The principal structural components of the offshore platform are the jacket, the piles, and the deck. The concept is very simple: the jacket is prefabricated on shore as a space frame, then it is transported to the site and seated on the seafloor. The piles are then driven through sleeves in the jacket, and connected to the sleeves. The deck is now set.

Jackets are also employed for offshore terminal construction, especially for the loading platform and breasting dolphins.

The typical offshore drilling and production platform does not exist for its own sake but rather is thought of as a necessary but expensive support for the primary functions which are the reason for the project. These functions are to drill wells, produce oil and gas, process it as necessary, and discharge it to pipelines to shore or a loading terminal. From the platform, conductors are installed, held by conductor guides bracketed out from the jacket. On the deck, derrick and drilling modules are installed, so that the wells can be drilled. Processing modules are installed on the deck, and all the necessary support modules for accomodations, power and water generation, sewage disposal, communication, and heliport. Cranes are installed to handle drill collars and casing, and all consumables from barges or supply boats to the deck. On the deck are stored drilling mud, cement, fresh water, and diesel oil. Other functions, such as re-injection of water or gas, may also be performed from the platform. An emergency flare stack is provided in order to flare excess gas. While diesel oil is used initially to fuel operations, produced gas may be used after production and processing are established.

The construction phase of a jacket for an offshore platform include fabrication, load-out, transport, launching, up-ending and seating, piling, deck installation, and module erection.

## The Steel Jackets

The typical jacket is subdivided so that the two narrowest sides are fabricated first, each laid out flat on the surface so that they may later be rolled up to position, the jacket itself being horizontal.

Cutting and fitting of the tubular intersections require precise work. In today's modern yard, this cutting is numerically programmed to ensure that the final weld gaps will be of the order of ± 3 mm. Fitting is done in the early morning when all the steel is at a uniform temperature. Rollup is accomplished by several large cranes, positioned so that they can initially lift the outside main leg, then walk toward the jacket centerline as the side is raised. This requires that the ground under the cranes have adequate support. Once vertical, or at the design tilt, sides are guyed off. Then individual cross members (bracing) which connect the lower and upper legs are set, fitted, and welded. Daily survey checks are run to prevent the development of cumulative errors.

Since the weight of the jacket has to be borne by the two lower legs or by similar longitudinal members attached to the bracing, additional vertical support is often required for this temporary condition of fabrication. Jackets that are large in plan and only moderately high may be fabricated in their upright condition. In this case, temporary runner beams are needed to support the jacket during fabrication and launching.

Large jackets often have complex intersection of tubular bracing, where 3 to as many as 13 braces intersect at a point, yet each must transfer the full force through the node. In this case, the nodes may

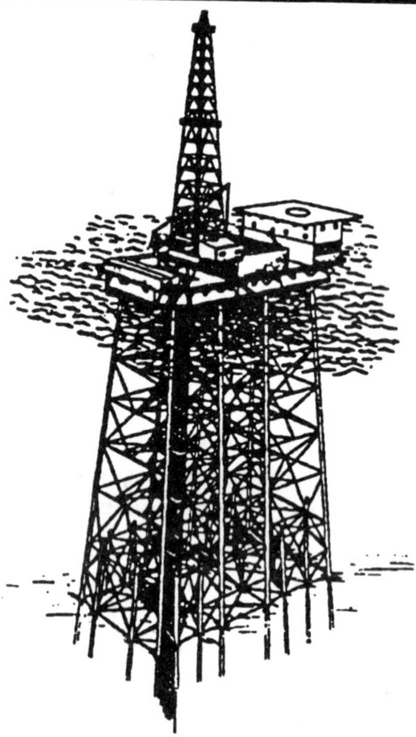

**Figure 12.1: Steel Jacket with Pin and Skirt Piles**

be fabricated first, detailed so that their connection to each brace is a right-angle joint, to be connected by a full-penetration butt weld. This same concept of prefabricated nodes is beneficial when a complex jacket is to be fabricated in a remote area of the world. Then the nodes can be shipped separately and the braces fitted at the site. For example, the brace may be beveled on one end, and allowed to run 150 to 300 mm long on the other end. At the final fabrication site it can be cut to fit.

All temporary attachments such as lifting eyes should be welded with the same procedures as the permanent members in order not to cause cracking or HAZ defects in the primary steel. Once their use is ended, these temporary attachments can be burned off 6 mm or so from the primary steel and then ground flush.

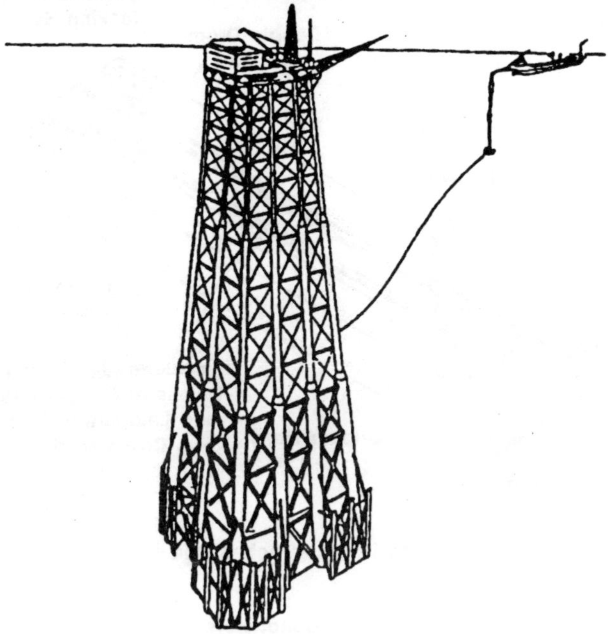

**Figure 12.2: Steel Jacket and Clustered Skirt Pile System**

All personnel must be fully aware of the catastrophe of the Alexander Kjelland semisubmersible *Floatel,* where a temporary entry, sealed with a substandard weld, later developed a fatigue crack in the main brace which resulted in the capsizing of the vessel with a large loss of life.

The jacket, having been fabricated on shore, must be transported to the site. Typically, it is skidded onto a launch barge. The launch barge is usually grounded at the dock, on a prepared and screeded sand pad. Water ballast is placed, sufficient to hold the empty barge on the bottom even at high tide.

For very heavy jackets and where the water is too deep to ground the barge or the tidal change too great, the barge must remain afloat. In this case the barge must be continuously ballasted to remain level and at the proper elevation relative to the skidways on shore, while the weight of the jacket is progressively transferred onto the barge. This can best be done by computer-controlled ballasting.

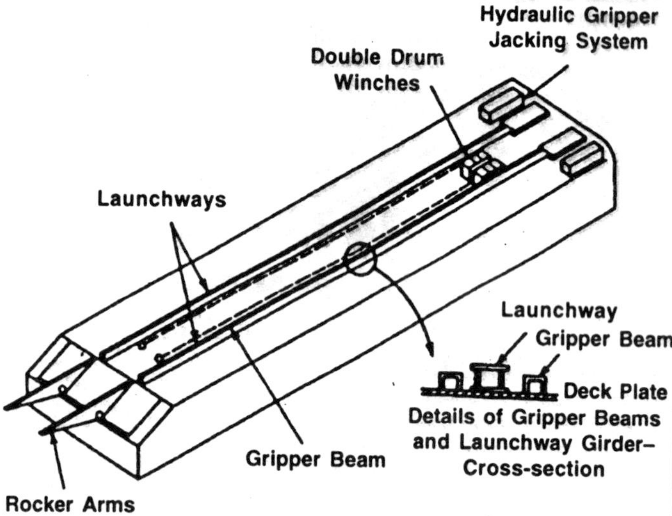


Figure 12.3: Launch Barge

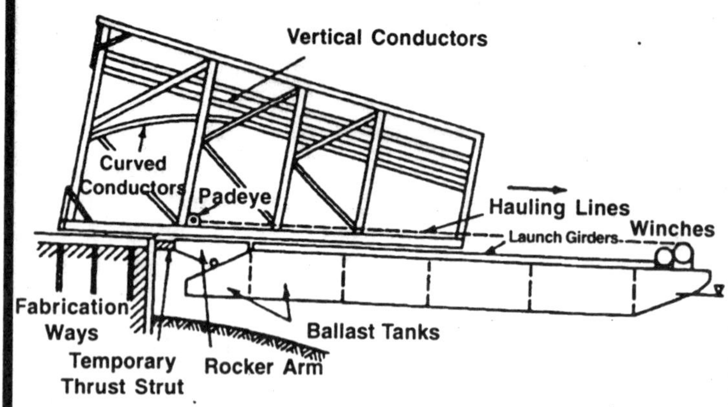


Figure 12.4: Load-out of Jacket onto Launch Barge

The barge must be secure against transverse movement induced by current, wind, or the wakes of passing boats. The alignment of the jacket must be accurately maintained while the jacket is pulled out onto the barge. For this reason, it is often preferable to ground the launch barge before load-out, whenever such factors as water depths make this practicable. Large modern launch barges are equipped with multiple ballast tanks and pumping capacity to enable the barge to be maintained at proper trim and draft.

During load-out, the jacket is supported on the fabrication ways, usually on two inner legs of the jacket. These are strengthened by plates to act like girders, able to support the jacket weight with some free span between points of contact. These girders are also converted into the bottom chord of a large truss, by using the basic platform bracing, often supplemented by additional diagonals, for example, to enable it to span between points of support, especially when part of the jacket is on the barge and part still on the fabrication ways.

Initial friction of the jacket on the ways may be as high as 10 per cent, especially if the jacket has been erected with its weight bearing on the ways continuously. In many cases, the initial fabrication is carried out slightly elevated above the ways by means of sand jacks. Alternatively, hydraulic jacks are used to permit removal of a filler piece. At time of launching, the jacket is lowered onto the skidways. To reduce the sliding friction, grease on hardwood, or heavy lubricating oil on steel, or even fiber-filled Teflon-faced pads, are used to reduce friction to as low as 1 per cent or less.

A check list of the operations relating to loadout of jackets follows:

1. Is jacket complete? Has the structure been analyzed for load-out stresses on the basis of the actual structure as fabricated at the time of launch?
2. Are the conductors, both straight and curved, in the same configuration and support condition as has been assumed in the analysis? Conductors, especially curved conductors, are often installed during the onshore fabrication and fixed to the jacket frame, as opposed to the vertical conductors which are often installed offshore, through conductor guides. Since decisions on the number, direction, and time of installation of conductors are often changed during the fabrication process, their support and tributary loads may differ from those used in earlier design.
3. Is the launch barge securely moored to the load-out dock, so that it will not move out during the loading? Is the barge properly moored against sideways movement?

4. If compression struts are used between the barge ways and fabrication ways, are they accurately aligned and supported so they will not kick out during launch?
5. Have the pull lines, shackles, and padeyes been inspected to ensure they are properly installed and cannot foul during load-out?
6. Is the barge properly ballasted? If the tide will vary during load-out, are ballasting arrangements made? Will ballast be adjusted as the weight of the jacket goes onto the barge? Are there proper controls?
7. If the ballast correction is to be made iteratively, step-by-step as the jacket is launched, are there clear paint marks so that each stop will be clearly identified?
8. If the load-out is taking place in an active or potentially active waterway, has the Coast Guard been asked to issue a Notice to Mariners to stop all traffic? Has a boat been stationed to stop the private power cruiser or tug which may not have received the Notice to Mariners?

Once the jacket is on the barge, the barge must be ballasted for sea. During load-out, many tanks will be partially full only, in order to control deck elevation and trim. Now with the jacket fully supported on the barge, these considerations are no longer active, and the tanks can be ballasted to suit the demands of the sea voyage.

Tanks should normally be either "pressed-up" full or else completely empty, to eliminate free surface and sloshing effects. The draft and freeboard will have been carefully selected to maximize stability and especially to prevent the outrigger-like legs of the jacket from dipping into the sea during roll of the barge. Trim will be adjusted to optimize tow speed and to give directional stability during tow; usually the barge will be trimmed down by the stern.

The above remarks apply when the barge has no restrictions from the load-out dock to the open sea. Many interesting variations arise in inland channels and bays, which have to be dealt with as special site-specific and jacket-specific operations. Examples follow:

1. Shallow water or a bar may limit draft and necessitate even trim.

2. A narrow channel may require that the overhanging legs of the jacket be high enough to clear dolphins, boat slips, even docks.
3. Fixed bridges may limit the height and necessitate ballasting down to deeper draft and occasionally to severe trim-down by the stern, even to the degree of submerging the stern of the barge. Since this reduces transverse stability (the water plane is reduced), this condition has to be checked with extreme care. This procedure was brilliantly executed in connection with the transport of platform Eureka under the Richmond-San Rafael Bridge in San Francisco Bay, with only 1 m clearance below the bridge deck girder.
4. The tides can be selected to give the greatest benefit at these critical stages. Tidal currents must also be taken into account, and adequate reserve tug capacity must be available to abort the tow and pull back if proper conditions are not maintained.
5. Wind from the side can cause the barge to heel; if the spread of the jacket legs at the bottom is 50 m, then even a 2° wind heel can cause a 1 m increase in height of the jacket leg, or perhaps a half-meter increase in draft. Once past these constraints, then the barge can be ballasted for sea.
6. The barge, while very stiff, is nevertheless a flexible member. The jacket is typically even stiffer than the barge. Therefore, adjusting of the ballast of the barge should preferably be done prior to tie-down for sea. If one scheme of ballasting was used for the inner channel tow and another will be used for sea, the tie-downs should be freed during the change in ballast to prevent imposing bending deformations on the jacket legs.

Tie-downs are installed after load-out and prior to entering the open sea. They are major structural systems, subjected to both static and cyclic dynamic loads. Therefore, the gravity and inertial forces involved must be calculated for all anticipated barge accelerations and angles of roll and pitch during the design storm adopted for the tow, usually the 10-year return storm for that season of the year and location. Since the loads are dynamic, impact must be minimized

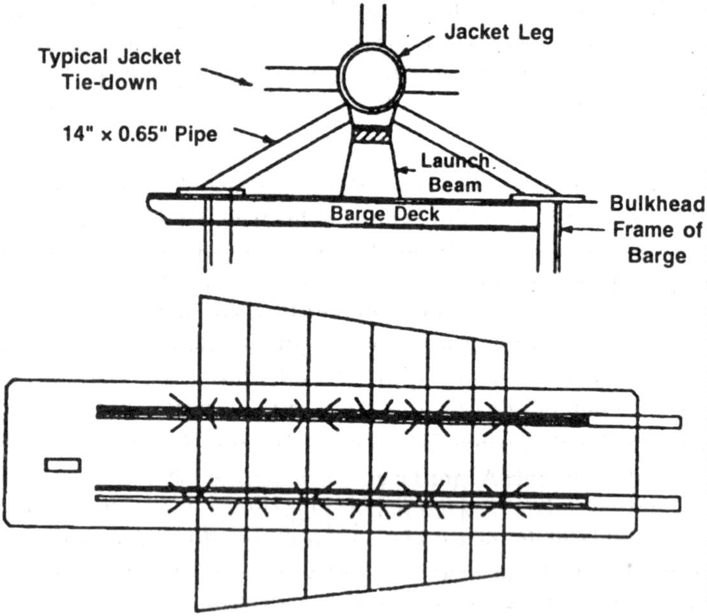


**Figure 12.5: Tie-down of Jacket on Barge**

and fatigue in a corrosive environment must be considered. The tie-downs will see approximately 14,000 cycles of fully reversing load for each day at sea. Fatigue has become a major concern in long transpacific tows.

Inertial forces are due to acceleration in heave, roll, and pitch and are therefore dependent on the period of response of the barge with the jacket loaded on board.

Shallow-water jackets, such as those employed for offshore terminals, are short and squat. They often may be loaded out and transported vertically rather than on their side. In this case, they may be skidded onto a barge, supported on temporary steel girders under or alongside the jacket legs. Since the weight of such jackets is usually less than 1000 tons, the load-out forces are not excessive. Therefore, because the launch beams on the jacket are temporary, they must be checked for possible eccentricity and also for web buckling; adequate lateral support must be provided. Once on board,

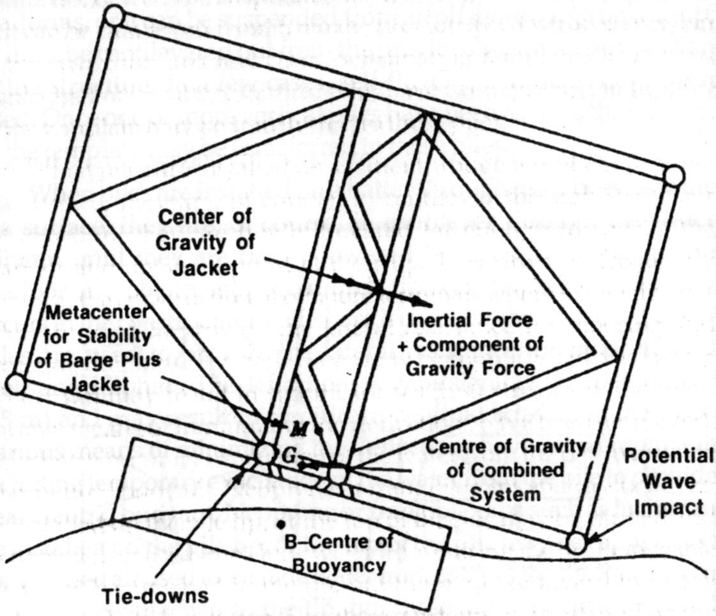


**Figure 12.6: Launch Barge with Jacket Under Heavy Roll During Storm**

the effects of barge response must be fully checked as to the loadings on the temporary girders, since the loadings will now have lateral components, and the webs of girders will no longer always be vertical.

The third method of load-out is that of the self-floater. In this case, the jacket is fabricated in a drydock or shallow basin. The legs on one side are typically made much larger in diameter to provide flotation for the entire jacket. Alternatively, extra legs or buoyancy tanks may be provided. Thus, for example, if in-service leg diameters of 2 m are required for structural purposes, the legs on one side may be 8 or even 10 m in diameter to provide the necessary buoyancy so the jacket can be self-floating.

This system has been successfully employed on several platforms off the California coast, on the Brent A, Thistle, Ninian South, and Magnus platforms in the North Sea, on the Maui A platform in New Zealand, and also on the Drift River Offshore Terminal in Alaska.

## Removal and Lifting of Jackets

Smaller jackets, designed for shallow water, are often lifted directly from the barge by one or two crane barges and set on the seafloor. The slings are attached and then the tie-downs and connections to the temporary skid beams are cut loose. Where long-period swells are being amplified and shortened by shallow water, significant differential movement may occur between crane barges and transport barge.

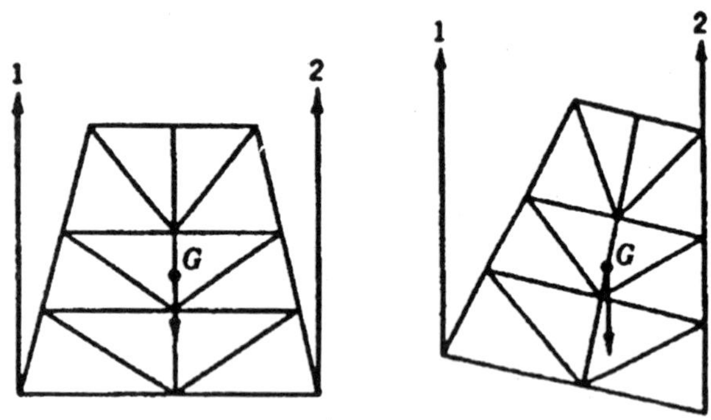


Case A–Two Crane Barges

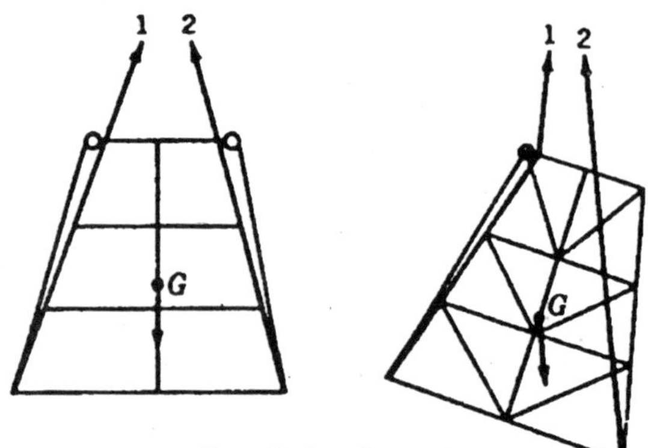


Case B–One Crane Barge

Figure 12.7: Lifting Off Jacket with Slings Attached Below Center of Gravity

Appropriate slack must be left in the lines during the period of cutting loose. The cutting-loose operation must be carefully pre-planned in order to prevent endangering the personnel, since most of the cuts must be made by hand. Short vertical guide posts may have been pre-installed at the load-out site to prevent lateral shifting of the jacket once it is cut loose. These braced vertical posts can form part of the tie-down frame; they must be adequately braced for impact. The jacket may also have chain stoppers acting as supplemental tie-downs. When the primary tie-downs are cut, the chain stoppers still hold the jacket laterally. These chains can then be severed remotely by power-actuated (explosive) cutters. Hydraulically-operated pins can also be pulled.

Slings for the jacket will preferably have been attached above the center of gravity of the jacket, so that the jacket will hang more or less vertically as it is lifted. In this case, it is only necessary to try to catch a group of lower swells or waves, and then hoist as rapidly as possible as the barge starts to rise on a crest. The dangerous time is the first wave crest after lifting off, when the jacket may once again be contacted by the barge or by the guide posts. These posts therefore should be only the minimum height necessary to prevent lateral displacement during cutting loose and have inclined protector plates welded across their top ends to minimize punching if the jacket leg should contact them on the second rise.

For a shallow-water jacket in the Gulf of Mexico, the jacket was skidded on its side onto the launch barge. At the site, it was lifted and set on the seafloor. Pre-attached slings then permitted re-rigging of the hook so that the jacket could be relifted from its end, causing it to rotate to the vertical for placement. Note again that certain slings must take the entire load of the jacket during rotation, which in turn reflects on the padeye design and the stresses in the jacket frame.

For any offshore jacket lifts, it will usually be found expedient to pre-attach the slings at the load-out site. Then when the barge arrives at the site, the eyes of the slings may be quickly raised by the whip line and placed over the horns of the hook, ready for the lift to take place. Tag lines must be used during the initial phase of lifting clear of the barge, to keep the jacket pulled slightly inward toward the lifting barge and thus preventing it from swinging.

Because of the heavy weight of a jacket and complications of such a lift, the crane barge(s) is usually pre-positioned and moored on site. The transport barge is brought in across the stern of the crane barge and secured. Then the jacket is lifted free. The transport barge is cut loose and pulled clear. Now the load is lowered to the seafloor. This minimizes the need for swinging either the barge or the boom, and keeps the crane barge picking over the stern where it has highest capacity and minimum roll response.

Most jackets are launched end-O from the transport (launch) barge. Jackets have been launched which weigh over 50,000 tons and which are over 400 m in length. This is one of the most dramatic operations in offshore construction, yet has been successfully performed many hundreds of times, just as ship launching has also proved successful. A few jackets have been damaged or even lost during launching, emphasizing the critical, dynamic nature of this operation.

The procedure itself is relatively straightforward. With relatively calm seas, the barge is headed into the sea. It is ballasted down by the stern so that it has an angle of 3° or more. The sea fastenings are cut loose. The jacket is then pulled off the stern by lines from the winches rigged around blocks at the stern and back to the bow of the barge. With larger jackets and dedicated launch barges, the jacket may be pushed off by hydraulically-operated gripper jacks. As the jacket moves end-O, off the stern of the barge, it finally reaches a point at which its center of load is beyond the pin of the rocker arms, The rocker arms then rotate to their limit (usually about 30°). The jacket now slides off the rocker arms into the sea.

There is a strong horizontal reaction imparted by the jacket to the barge, causing the barge to surge forward, at the same time as the stern kicks ups due to the release of the jacket load. If this is a manned operation, the personnel, stationed near the bow, must have a safety line to avoid being thrown off the barge by this rather violent reaction of the barge. In most modern cases, this operation is carried out by remote control, unmanned, to avoid the danger. An umbilical cord from the tug or radio may be used to actuate the launching system.

The jacket, leaving the barge, has combined downward and rotational momentum. It will therefore usually plunge, with some jackets plunging even deeper than nominal diagonal length, before slowly returning back to sea level in a horizontal attitude. Most jackets are designed to ride, self-floating, on the upper side legs,

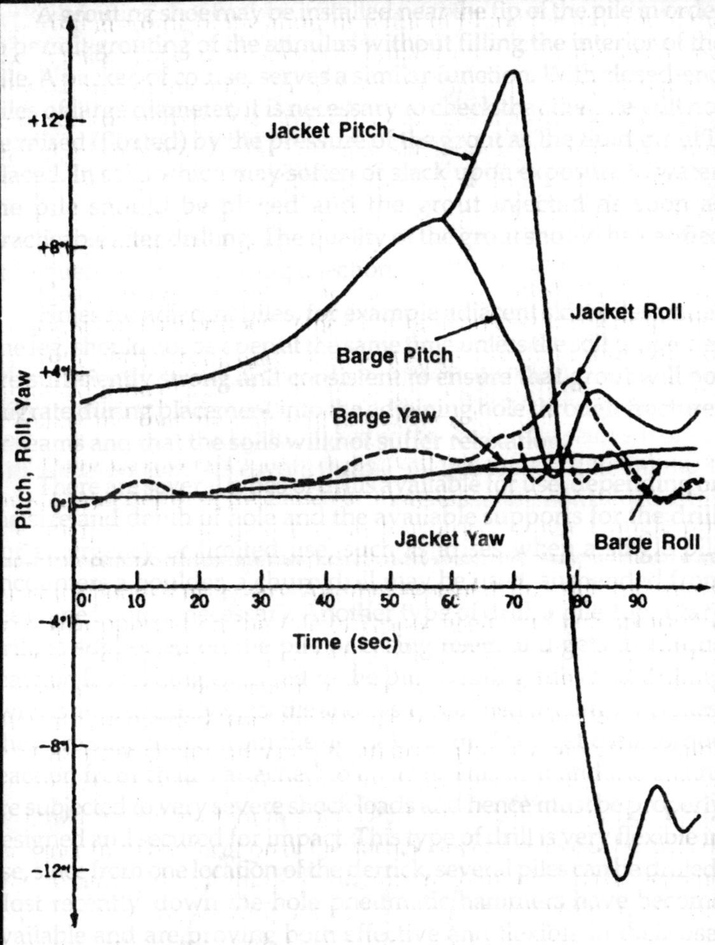


**Figure 12.8: Motion Responses During La··nch of Large Jacket**

with these about half immersed. This means a freeboard of only half the diameter of a jacket leg.

To return to the launching operation, starting friction may be relatively high. This will require the use of high pulling forces or thrust from jacks, opposite in direction to th·se applied in loading the jacket. As the jacket moves down the laun.:hing ways, its weight is imposed progressively on a smaller and smaller length of the two central jacket legs, until finally all the load is that at the rocker arm. The jacket now rotates partially into the water so that it is supported over two zones: the water and the rocker arm. The jacket continues to slide, the two legs still carrying a high portion of the total load until the jacket finally slides free. Thus the jacket legs will normally require reinforcement to take the bending and local concentrations of load. Note that while the vertical load at the time the rocker arm rotates will normally be the maximum, there is in addition a friction force acting parallel to the jacket leg.

The worst thing that can happen during a launch is for the jacket to skew sidewise and thus not only tilt the barge to cause the jacket to roll but also cause loads on the jacket frame at points and in amounts for which it is not designed. The proclivity to roll is in part due to the raise of the bow of the barge out of the water as the center of gravity of the load, *i.e.*, the jacket, moves aft, thus reducing the water plane moment of inertia.

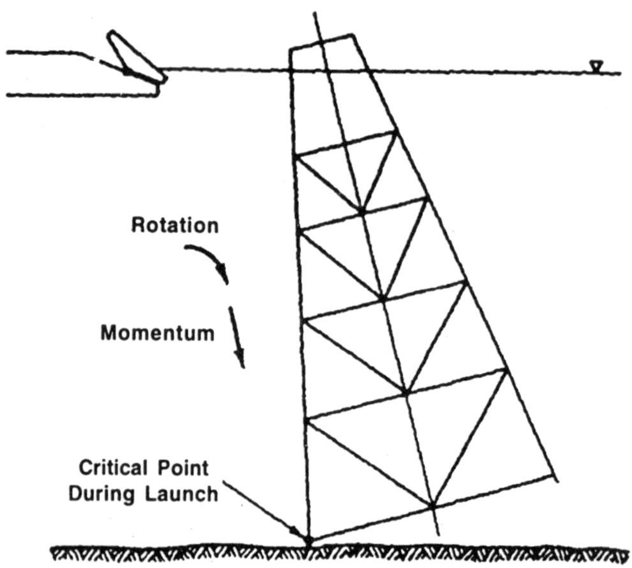


**Figure 12.9: Potential Impact of Jacket on Seafloor**

During initial movement of the jacket on the barge, moving astern until its center of gravity reaches the rocker arm pin, the jacket can be kept properly aligned by controlling the jacking/pulling system and by steel plate side guides on the two center legs of the jacket, which hold them on the launching ways. Feedback instrumentation must be installed to verify that the jacket is correctly aligned as it reaches the rocker arms.

As the jacket rotates into the water, there are impact forces (similar to wave slam) on the legs and cross members and any temporary buoyancy elements, tending to tear them loose. There are also inertial forces acting on any piles which were pre-installed in the jacket legs or sleeves, tending to cause them to plunge downward. In the case of the Magnus platform in the North Sea, several piles ruptured the end closure of the jacket legs. The piles plunged to the seafloor and were severely damaged.

Jackets have been loaded and launched with the lower end (base) launched first, and also the reverse, where the top is launched first. Present practice appears to favor launching with the top of the jacket first.

Many of the tubular members of the jacket will have been subdivided to be watertight and empty, in order to provide the needed buoyancy to cause the structure to float properly. These are subjected to hydrostatic forces, principally hoop stresses but also complicated by axial compression due to hydrostatic force on the end. Supplemental hoop reinforcement may be needed in order to resist the combined stresses. The tubular members and temporary buoyancy tanks may also experience ovaling forces due to drag as the water rushes past the bracing and legs during launching. The design against buckling must consider initial out-of-roundness of the tubulars.

The jacket must be launched in sufficient depth of water so that there is no danger of it hitting the bottom, again taking into account the momentum of launch and the diagonal length across the jacket. On several occasions, jacket legs have been damaged by hitting the seafloor.

## Installation of Self-floating Jacket

The up-ending of smaller jackets has been often accomplished by a combination of differential ballasting, augmented by the lift

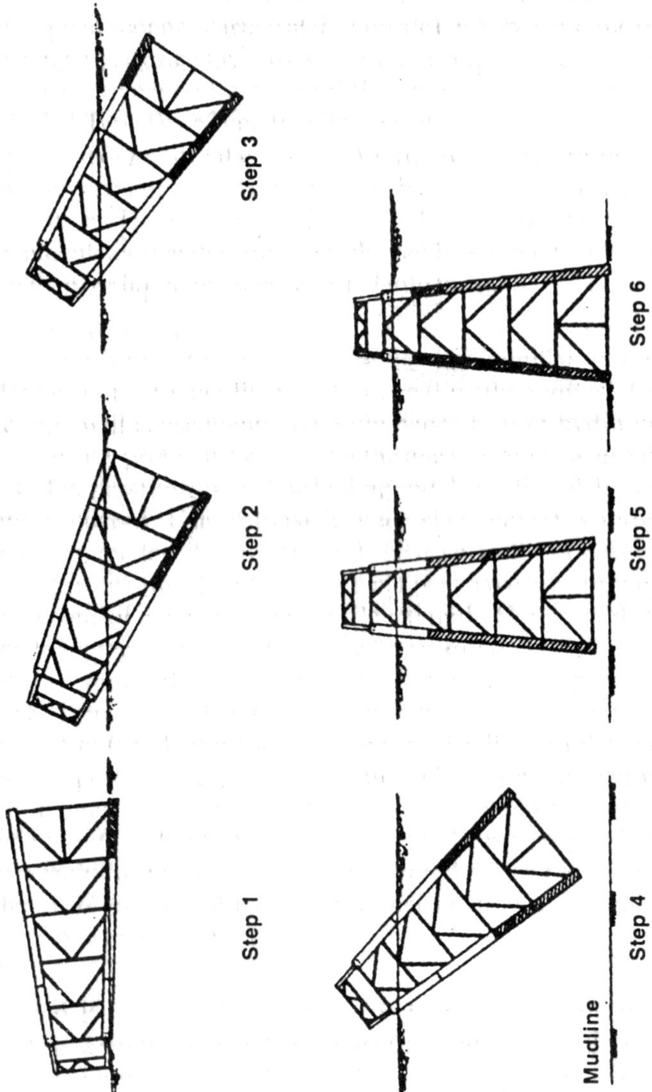


Figure 12.10: Installation of Self-floating Jacket

from the crane boom of an offshore derrick barge. Although this provides excellent control, it involves several potentially dangerous dynamic aspects.

First, the jacket, having a large actual mass plus an added mass (hydrodynamic mass) of almost equal magnitude, cannot respond to the accelerations induced in the boom tip by the heave and pitch of the barge. These latter have a typical double-amplitude period of 6 s, which means that it is the boom and derrick barge that are pulled down when the wave crest passes, rather than the jacket being pulled up.

There is, of course, elastic stretch in the wire rope falls; hence use of as many parts as practicable is desirable. There is also the flexibility in the boom and stretch in the topping lift lines. Nevertheless, this procedure is safe only in a very calm sea. The slings for this up-ending should have been pre-attached, to be readily accessible above water, for hooking on.

The crane boom can provide control of the jacket attitude; but the primary up-ending moment must come from differential ballasting in which water is flooded into the lower portions of the jacket legs on the high side. As the jacket rotates, water may be drained out of upper bracing.

API RP2A provides in part: "Generally, the up-ending process is accomplished by a combination of a derrick barge and controlled or selective flooding system. This up-ending phase requires advance planning to pre-determine the simultaneous lifting and controlled flooding steps necessary to set the structure on site. Closure devices, lifting connections, etc. should be provided where necessary. The flooding system should be designed to withstand the water pressures which will be encountered during the lifting process."

Large jackets have extensive ballasting and control systems installed, to permit flooding and venting, as well as hydraulic lines with which to operate valves. No cranes are used for the large jackets, there is too much danger of overload.

The bending moments and forces induced in the jacket during up-ending must be determined, in order to prevent overstress in the jacket frame. Any tubular members which are empty or partially empty during the up-ending process must be able to withstand the combined hoop forces and axial forces induced by the water pressures at the depths involved; these conditions and forces may not necessarily be the same as those during launching or in service. Failure to recognize the effect of combined stresses is believed to have been partially responsible for the collapse of the temporary buoyancy tubes on the Frigg DPI platform, which resulted in loss of the jacket. Note especially that self-floating jackets will first experience significant hydrostatic pressures during up-ending.

One means of countering high hydrostatic pressures is through internal pressurization with compressed air. On the BP Forties platforms, nitrogen gas, released from liquid nitrogen, was used to internally pressurize the temporary buoyancy tanks.

Whenever possible, the use of internal balancing pressurization should be avoided due to the constraints upon design and handling that it produces. If it is used, the following should be noted.

1. The rate of pressurization should not exceed the structure's ability to withstand stresses induced by the increased temperature due to compression of internal air.
2. The process has to be capable of being arrested at any stage without the need for power.
3. Note that the expanded gas from a liquid, *e.g.*, liquid nitrogen, is extremely cold and may freeze valves. Compressed air, on the other hand, can get very hot and interfere with controls and computers.

Control of relatively small jackets has usually been by umbilical (electric-hydraulic) from the derrick barge, actuating the opening and closure of valves, and feeding back information on progress of flooding. Usually the valves are equipped with spring closures to automatically close in event of power or hydraulic failure. Screens are provided over intakes to prevent entry of debris which might prevent closure of valves.

Pressure sensors, sensing the rise in pressure of the air compressed at the top of a member as it is flooded, or of the water at the bottom, provide necessary information. As more-sophisticated and larger jackets are installed, valve-position indicators may also send the signals back to the control station on the barge.

As jackets have become larger, the up-ending process has usually been carried out remotely, without involvement of the derrick barge for lifting control. Three-legged jackets usually roll during up-ending, making it unsafe to have a line from the boom, while deep-water jackets usually traverse too great an arc for the boom to follow. Remote control has been exercised as before, through an umbilical. However, umbilicals have been broken by the extended sweep of the upper end of the jacket as it rotates. Radio control has thus been found more reliable and is now the state-of-the-art for major jackets.

As a backup, there is usually a station on the upper end of the jacket, where manned controls can be activated in an emergency. The personnel are usually not on board the jacket during the initial part of the up-ending but may be transferred later by helicopter or boat. For this latter purpose, a rope ladder is arranged to hang down from the control station.

Large jackets, designed for deep water, obviously require a more-sophisticated plan for up-ending in order to avoid overstressing of the jacket frame. The large legs of self-floaters can be subdivided both in plan and length. Similar subdivision can be carried out for those jackets in which only skirt piles are employed, thus permitting the legs to be divided by transverse closures. Large legs and temporary buoyancy tanks may be pressurized internally to resist hydrostatic pressure.

Up-ending is usually planned by means of a computer program which takes into account the constantly changing configuration of submergence and the changes imposed by ballasting. Once a suitable plan has been developed, physical model tests are run. These serve two purposes: first, to verify the behavior of the jackets during up-ending and, second, to acquaint and train the key people–barge superintendent and offshore engineer–in this complex dynamic operation.

After up-ending, the jacket, now in vertical attitude, and with a draft only 3 to 5 m less than that at the installation location, is towed slowly to its final site location. Wherever feasible, the up-ending is, of course, carried out at or in the immediate proximity of the final site. However, the diagonal depth of the jacket may exceed the final draft. Where seafloors are very uniform in depth over a large extent, as in parts of the North Sea, this may necessitate up-ending some distance from the site and then towing it to final location. Such final tows will have a bridle pre-attached near the center of rotation of the jacket, so that the jacket will remain vertical in the final tow. Towing force and speed is purposely reduced to a minimum. To eliminate or reduce this extra step (final positioning tow), the present trend is to use temporary buoyancy tanks which will enable up-righting near the final site.

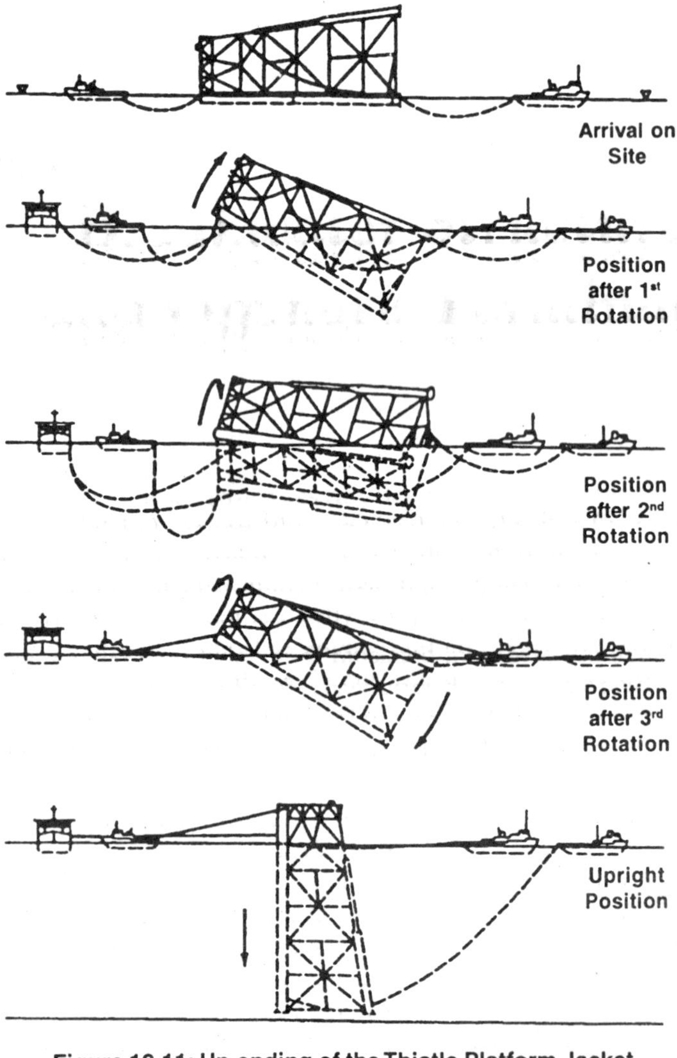


Figure 12.11: Up-ending of the Thistle Platform Jacket. Ballasting Sequence not shown.

## The Up-ending of Self-floating Platform

To ensure that the jacket will be installed in its proper location, an offshore derrick barge is normally moored on location. In shallow water, this mooring is accomplished by the derrick's own anchoring system, with the anchors being carried out by anchor-handling boats. Once set, a pull is taken successively on each anchor line to ensure the anchor is properly seated. The final location and orientation of the derrick barge is then established by means of survey, principally DGPS and electronic survey, but often keyed in to any preset acoustic transponders on the seafloor.

API RP2A requires that the anchor lines be of sufficient length for the water depth at the site and that the anchors and lines be of

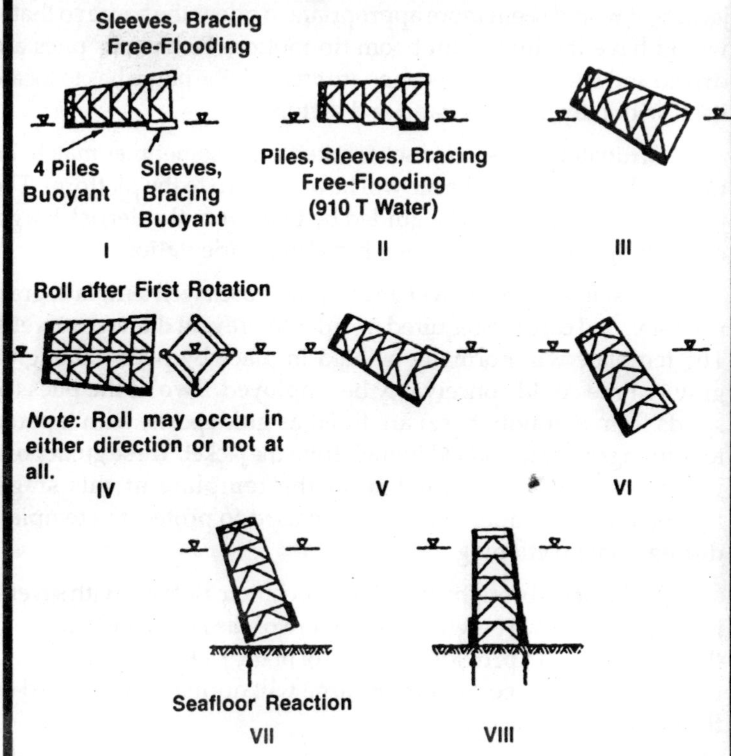


**Figure 12.12: Steps in the Up-ending of Self-floating Platform**

the proper size (weight) and shape to hold against the maximum combination of wind, current, and waves.

In deep water, the derrick barge's own anchoring system may have inadequate length of lines and hence mooring buoys are pre-set to which the derrick barge lines can then be run.

API RP2A also contains a rather curious section suggesting that, where holding ground is poor or the mooring system cannot be made fully adequate, the derrick barge should be located so that if the anchors do slip, the barge will move away from the platform. This provision may be appropriate for small platforms being constructed with marginal equipment, or may have been intended primarily for application at later stages of construction when the jacket is firmly seated in place. However, for installation of major jackets, it would seem more appropriate to orient the barge so that it would have the minimum boom tip motion. Further, as piles are driven and later, as deck sections are erected, the barge has to locate itself within the limiting radii and sectors.

Fortunately, the second set of criteria will sometimes match the first, in that the derrick barge will have its stern to the platform. The jacket location will then be guided by lines from the derrick barge, controlling not only its location but also its orientation.

To install a platform over an existing subsea well template, great precision and care are required in order to prevent damage to wells. The template will normally be held in place by piles although a gravity base could conceivably be employed. Two of the piles (or spuds from a gravity base) are fitted as guideposts, with tapered tops to engage cone-shaped funnels from the jacket. These guideposts will usually be decoupled from the template at this stage. Independent "bumper" piles may be used to protect the template during final positioning.

The jacket is brought into proper position, floating with several meters of clearance between the bottom of the jacket and the top of the guideposts. To provide full control of the jacket position during this operation, a second derrick barge will normally be moored on the far side of the jacket.

In early development of this technique, guide lines from the jacket top were attached and tensioned to give a visual indication of location and verticality. Today, with sophisticated sonar (acoustic) locators and transponders, plus video cameras and inclinometers which can be mounted on the legs, it is feasible to dispense with the guide lines and place full reliance on the instrumentation. Redundancy in instruments must, of course, be provided in the event of malfunction of any instrument. The jacket is now slowly ballasted down to engage the guideposts, and then on down to seafloor contact with the mud mats.

Whether set independently on the seafloor or over a well template, the jacket now must derive temporary support from the mud mats bearing on the soils at or just below the surface. The jacket must be self-supporting until pin piles can be driven. It is important that the jacket be level and remain so within a small tolerance until the piles are installed. The effective weight of the jacket on the bottom may be controlled by ballasting. This permits moderate adjustment of level of the jacket, which may be supplemented by moment induced by lines from the controlling derrick barges.

For large jackets, where precise leveling is required jacking devices can be built into the connection between jacket legs and the mud mats. Commercially, available pile-supported leveling systems are available. These hydraulic leveling tools work in conjunction with temporary gripping devices which hold the jacket in position while pile-to-jacket connections are being completed.

The jacket must be approximately level when the piles are driven to avoid introducing unacceptably high bending stresses in the piles. Thus leveling after piles are installed must be limited to relatively minor corrections. A careful evaluation must be made of the soil loadings during this phase. The jacket will be bearing at this stage either on the bottom bracing or on mud mats or a combination of both. The weight of the jacket must include any piles or conductors which are being supported by the jacket during the installation.

The bearing pressure on the soil must be within allowable limits under the combination of direct load and that due to waves and current during the piling phase. API RP2A allows a one-third increase in allowable soil-bearing values during this phase if wave action is considered. This may be roughly acceptable in smaller installations. A much more thorough analysis is required for major structures, taking into account short-term consolidation settlements, and the effect of cyclic lateral and vertical strains. Scour around and under the mud mats must be prevented. This may require filter fabric and stone placement.

All structural elements bearing on the soil or supporting the mud mats must be adequate for the maximum bearing loads anticipated, including those due to storm. The design of the mud mats should also address the failure mode, to be sure that any structural failure will take place in the mat proper, rather than by damage to the permanent jacket legs or braces.

Mud mats were originally timber planks affixed to the bottom bracing to increase the bearing area. With major jackets, these mud mats are now structural steel, heavily reinforced flat plates, carefully designed to provide proper bearing. They are frequently tailored to fit the bottom contours; in the case of the Hondo platform off the Southern California coast, there was 20 m difference in elevation from the deepest to the shortest leg. This means that the jacket must be accurately oriented as well as positioned.

However, is more difficult to penetrate because of its large area. Jets can be pre-installed, with nozzles acting along the underside of horizontal bracing to wash out material from under the bracing and lubricate the sides.

For self-floating jackets, which typically have two enlarged legs, or where pile sleeves are of large diameter, supplemental means may also be necessary to cause them to penetrate. Jets can be arranged inside to break up the plug and an airlift or eductor system employed to remove the material. These can be designed to operate below the pile closure.

For the Maui A platform off New Zealand, jet and airlift systems were built into the two enlarged legs, to enable the material within the large-diameter legs to be progressively removed.

In poor soils, one other way around the dilemma of providing vertical and lateral support to the jacket during this early phase is by driving four temporary piles to a short penetration only. The jacket can then be leveled by jacking, lifting, or ballasting, and temporarily welded off to these four piles

The "temporary piles" may be four of the permanent piles, driven initially only to a small penetration. They would typically have been transported with the jacket, to expedite their release and installation.

Sometimes these short piles are made permanent and used as spuds for lateral support. In most cases, after the remaining permanent piles have been driven, these temporary piles are cut loose, and raised as necessary to release any bending stresses. Add-ons are welded on and the lengthened piles are now driven to final penetration.

## The Deck Sections

API RP2A requires that the deck elevation be within plus or minus 75 mm from the design elevation and shall be level. The degree of level is usually limited to about 300 mm differential height across the longest dimension of the platform, but in any event should ensure proper drainage and proper operation of processing equipment.

Deck sections are now to be lifted on. With smaller platform, the "pancake" concept was often adopted, in which some of the permanent equipment was pre-attached to decks, with each deck of the platform being lifted on in succession. After each deck was erected, remaining equipment for that deck was set.

With larger platforms, the "deck" now consists basically of module structural support frames, consisting of girders and trusses onto which large modules of assembled and integrated equipment are set. The initial sections have legs extending below them, with stabbing guides to fit into the piles or jacket legs. The stabbing guides are so configured that they also act as backup plates. Since the mating leg is the same diameter and wall thickness as the extension to which it is to be joined, a full penetration girth weld is made, similar to the splice in a pile.

To aid in stabbing four legs into four sleeves, the stabbing guides may be made slightly different in length so that one can be entered first, then the module rotated so the second can be entered and then the whole lowered to fit the remaining joints.

Transport of a large deck module will usually be by barge, although smaller modules and equipment may be transported by supply boat or on the derrick barge itself. The weight of modules has grown in recent years, to 500 tons, then 1000, and most recently, over 10,000 tons. Mammoth derrick barges are now available to lift this weight and more. The purpose is to enable more complete assembly onshore.

Piles which extend on through legs of the jacket to the deck can also be secured to the jacket by welding; this system has been much used in the past where piles extended up above water. It is also currently employed in offshore terminal construction where jackets and pin piles are employed. The transfer of high cyclic axial loads from the top of the pile into the jacket leg requires careful consideration of weld details, since the welding will have to be carried out under adverse conditions of wetness (spray), perhaps low temperature, and while the jacket may be vibrating under wave action. Grouting helps to reduce vibration and limit excessively high stresses at the welds.

Steel shims are used to center the pile in the jacket leg; these are usually one quarter or one third segments of steel pipe of the proper radius. The welds are best designed as shear welds, from the pile to the shims, then from the shims to the jacket leg.

The lateral resistance of the installed platform is developed by the $P/y$ (lateral load-deflection) of the pile-soil system, which in most soils takes place over the top five diameters of the pile. Since this is normally the zone of weakest soils, lateral resistance may be critical. Several schemes to enhance the properties of the soil in this

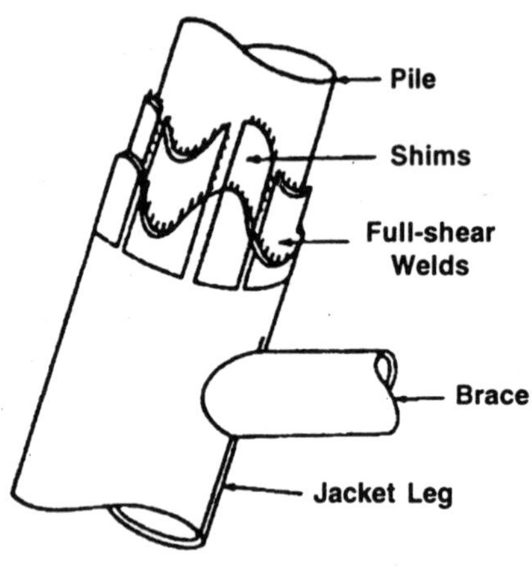


**Figure 12.13: Scalloped Welded Connection for Transfer of Load Between Pile and Jacket**

zone have been proposed, including built-in drainage systems at the tip of the jacket legs (or sleeves). To prevent annular gaps from forming around the pile under cyclic wave loads, pea gravel has been dumped on the seafloor around the pile: as a gap opens on one side, the pea gravel works down and wedges the pile, preventing progressively increasing displacements.

For the same reasons, it is important to prevent scour around the piles and bottom of the jacket, both when the jacket is temporarily supported by the mud mats and in service. Some prospective areas for platforms, for example, Sable Island off Nova Scotia, have sandy seafloors and high bottom currents, and have shown rapid scour around the legs of jack-up drilling platforms. Shallow-water areas, with sandy bottoms, where wave action may be severe, are especially suspect, since scour due to eddy action may be augmented by the pumping action of the jacket vibrating and rocking under the waves.

Scour protection around jacket legs can probably be most expeditiously and practically accomplished by the placement of graded rock through a long tremie pipe. Obviously, the depth of practicability is limited, but fortunately so is the depth at which scour action usually occurs. Alternatively, controlled dumping from the surface has been utilized with generally satisfactory results.

Conductors are now installed, in much the same manner as piles. The lower section of some of the conductors may have been carried out with the jacket, but for the most part they are transported by barge, threaded in through the conductor guides, extended by add-ons, and driven to the required penetration. Since they are usually of smaller diameter than piles, that is, about 750 mm diameter × 25 mm walls, and usually penetrate to less depth than the piles, they are easier to drive, and smaller hammers can be used. Their penetration requirement is determined primarily by the ability to seal off flow during drilling, so that drilling mud will not escape to the sea. They must also be driven to a sufficient depth to prevent escape of shallow gas, which could form a flow path for future release. Conductors also must provide vertical support to the wells. Conductors may be installed by the drilling rig, which may use either a pile hammer or drilling and jetting techniques. In mudslide areas, the conductors may be enclosed within a larger-diameter tubular, which provides the strength and stiffness to resist the lateral forces from the moving mass of mud.

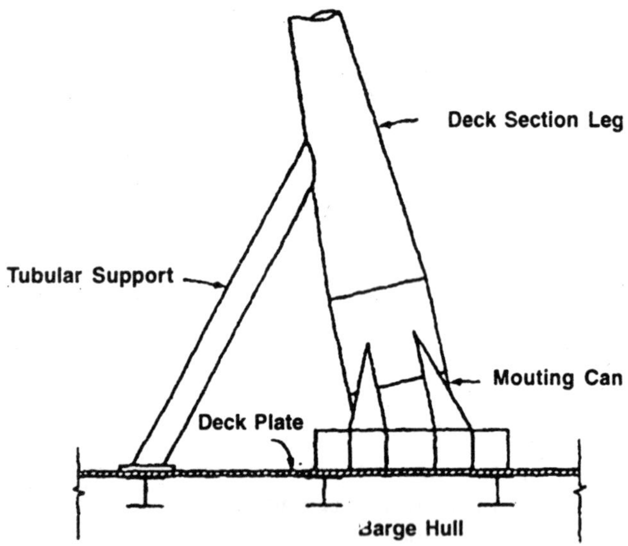


**Figure 12.14: Sea-fastening for Leg of Deck Section**

The latest generation of heavy-lift derrick barges is of the semisubmersible type. This, of course, is a trade-off between the reduced motion response of the barge and the concomitant reduction in stability, especially in roll, as the load is lifted. For this reason, heavy lifts are generally made over the stern, with the swing being used only for minor adjustments in position to engage the stabbing guide. In fact, for heavy lifts such as these, a sheer-legs crane barge is also suitable, with the minor positioning adjustments being made by the deck engines. The largest offshore derrick barges in the North Sea now are fitted with two huge cranes, one on each stern quarter, so that their combined capacity may be used. For heavy lifts, boom tip response is very critical. Onboard minicomputer programs have been developed to optimize the heading and boom angle.

The derrick barge is pulled back from the platform, and the cargo barge, with the large deck unit on deck, is pulled in across the stern. The lift is made and the cargo barged pulled clear. The derrick barge now pulls astern to the platform, where it sets the deck unit.

Because of the numerous parts of line needed for such heavy lifts, getting rid of the load after having landed it on deck is often a problem, even with free overhaul release. The stabbing guides must be designed to remain engaged, once entered, since otherwise on the next heave cycle, they may disengage.

Load-out and transport of these large deck units and modules on a cargo barge requires procedures and sea fastenings similar to those of the jacket, with added complications. The unit, with its four or more legs extending downward, is difficult to support. Large reinforced plate-bracket assemblies are needed to distribute the load (static plus dynamic) over the deck so the leg will not punch through, and can be supported both on the skidway and on the barge.

Decks have thus undergone a series of major evolutionary developments, spurred by the recognition that the greatest portion of the total costs of an offshore platform is generally in the processing and support equipment and the greatest labor demand is in hookup and testing. Operating with high volumes (50,000 to 300,000 barrels of oil per day), high pressures, and high volumes of gas requires precision assembly and thorough testing. Hookup is a major demand on skilled manpower and on a large platform can require 2 to 2.5 million man-hours. Significant savings in cost and time can be achieved by carrying out this work under favorable conditions at inshore shipyards.

At the same time, the total deck "payload" has grown from 7000 tons to over 50,000 tons, partly because of increased requirements such as gas reinjection, and water flooding, partly because of more remote and demanding environments requiring weather protection which in turn requires more ventilation, large helicopter services, and greatly enlarged quarters and support. The typical jacket-pile structure is very sensitive to total deck load. One way to reduce the total is by integrating the deck or at least the modules to make more effective use of the deck structure. The evolution then has been from individual deck sections and individual pieces of equipment, to module support frames and large integrated modules, to completely integrated decks.

*Chapter 13*

# The Nature of Deep Sea Constructions

## Introduction

In the offshore construction industry, the deep sea construction is the most important task.

The discovery of giant fields for oil and gas in deep water has presented a major challenge to the industry, resulting in remarkable developments in the way of equipment, procedures, instrumentation, and remote operations.

What constitutes the deep sea? When international agreement was reached on national jurisdiction, 200 m was considered the limit beyond which development of resources would be prohibitively costly and beyond the capabilities of technology. When the first edition of this book was written in 1986, the demarcation had risen to 300 m and the term "Deep Sea" was being applied to platforms and pipelines in 500 m. Today, at the close of the century, bottom-supported structures are being constructed in over 1000 m water depth, and subsea operations, supported by floating FPSO spars, TLPs, and drilling vessels, are in place in 1600 m. ROVs capable of 2500 m and even 3000 m are now available. Drill ships are being constructed to work in 3000 m, which is almost the maximum depth of the Gulf of Mexico.

Pipelines, flow lines, and risers have been installed to subsea completions such as the Mensa project (1700 m) and are planned for the Exxon Spar project in 1600 m.

Military activities, such as the deployment of acoustic sensors for recovery of armament and equipment, including retrieval of parts of a Soviet submarine, have been carried out at depths up to 6000 m. Test facilities for ocean thermal energy conversion (OTEC) have included pipelines suspended to a depth of 600 m. Tests of manganese nodule mining equipment have been conducted at a depth of 2000 m.

The deep-sea frontier is rapidly emerging as an area of great activity. Exploratory drilling for oil has already been carried out at depths of over 3000 m. The Deep Sea Drilling Project included the successful drilling and reentry of a hole at a depth of 6000 m. Potential exploitation of the polysulfide mineral deposits from midocean rifts will require specialized dredging operations, with equipment and materials capable of operating in hot brine. Manganese nodules are concentrated on plateaus and basins lying at 2000 to 4000 m depth, requiring efficient dredging systems capable of operating at such depths. OTEC systems are generally based on the utilization of the cold water from 1000 m depth. The floating structures for this concept may require mooring in 4000 m. The deployment of sensor devices with cable moorings and of large surface and subsurface buoys has been carried out throughout almost the entire range of ocean depths.

For the deep sea operations following things are considered:

1. Extreme hydrostatic pressures.
2. Density changes in liquids, including seawater, due to high pressure and low temperature.
3. Reduction in volume of solids due to bulk modulus effects (usually important only for low-modulus materials such as polyurethane foam).
4. Absorption of water into concrete and other solids.
5. Absorption of gases into solids.
6. Miscibility of water and other fluids.
7. Change in strength of materials due to high triaxial stress states.

8. Density and other currents–at depths of 1000 m the currents may be of the following order: density currents: 0.2 to 0.5 knots; Internal wave-generated currents: up to 0.6 knots; tsunami currents: up to 0.6 knots.

   Currents may produce vortex shedding and thus require the installation of "spoilers" which may also produce dynamic responses in long risers and strumming vibrations in long cables.
9. Internal waves.
10. Density layers (stratification) of the water column.
11. Leaks in seals of hydraulic systems, electrical connectors, etc. due to high pressure.
12. Difficulty of control as a result of time lag in response of hydraulic systems due to the long length of lines. For this reason, deep-sea well controls use electro-hydraulic control.

## The Compliant Structures in the Deep Sea

At depths beyond 300 m, there is a trend to the use of compliant structures whose period is significantly longer than that of the design wave. These are structures which are intentionally flexible transversely, yet fixed to the seafloor for shear and axial support. Several concepts have been developed to achieve this response: the guyed tower, the free-standing flexible tower, and the articulated column. The towers themselves are typically trussed columns with a relatively constant cross section, although a large-diameter tube is also conceptually feasible. The base is either spudded or piled, in order to support the tower and provide restraint against lateral shear.

## The Guyed-tower Concept

Fabrication of the tower is carried out on a ways, as with any jacket. Since this structure is primarily for deep-water use, it will often be fabricated in two halves, just as was done for the Hondo platform. Preferably the two halves will be built as one, then later separated, to ensure perfect match.

However, if sufficient yard space is not available, a short section at the juncture, incorporating both mating sections, can be constructed first. Then this section is skidded to the far inshore end

**Table 13.1: Properties of Typical Fiber Ropes**

| Size | | Nystron Braid[a] (Nylon/Polyester) | | | Nylon/Multifilament Polypropylene Rope[b] | | | Polyester Rope[c] | | | Kevlar[d] | | |
|---|---|---|---|---|---|---|---|---|---|---|---|---|---|
| Diameter (in.) | Circumference (in.) | Weight per 100 ft (lb) | Minimum Breaking Strength (lb) | Elastic Elongation at 25% of Breaking Strength (%) | Weight per 100 ft (lb) | Minimum Breaking Strength (lb) | Elastic Elongation at 25% of Breaking Strength (%) | Weight per 100 ft (lb) | Minimum Breaking Strength (lb) | Elastic Elongation at 25% of Breaking Strength (%) | Weight per 100 ft (lb) | Minimum Breaking Strength (lb) | Elastic Elongation at 25% of Breaking Strength (%) |
| 2 | 6 | 114 | 121,000 | 7 | 93 | 88,400 | 7 | 124 | 105,400 | 3 | 132 | 172,000 | 1 |
| 2¼ | 7 | | | | | | | | | | 180 | 224,000 | 1 |
| 3 | 9 | 268 | 272,000 | 7 | 210 | 193,000 | 7 | 294 | 236,300 | 3 | – | + | – |
| 4 | 12 | 470 | 460,000 | 7 | 371 | 329,000 | 7 | 515 | 399,500 | 3 | – | – | – |
| 5 | 15 | 719 | 683,000 | 7 | 590 | 505,000 | 7 | 788 | 593,300 | 3 | – | + | – |
| 6 | 18 | 988 | 921,000 | 7 | 836 | 698,000 | 7 | 985 | 731,850 | 3 | – | + | – |
| 7 | 21 | 1,348 | 1,233,000 | 7 | 1,080 | 884,000 | 7 | 1,478 | 1.071,850 | 3 | – | – | – |

[a] *E* at 25 per cent B.S. = 140,000 psi.

[b] This rope has neutral buoyance! *E* at 25 per cent B.S. = 100,000 psi.

[c] *E* at 25 per cent B.S. = 280,000 psi.

[d] *E* at 25 per cent B.S. = 1,400,000 psi.

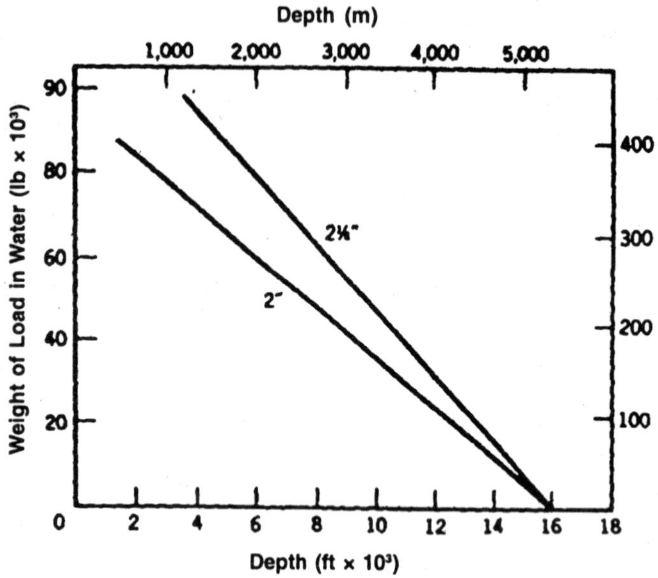


**Figure 31.1: Load-carrying Capability of 6 × 41 Fiber-core Wire Rope**

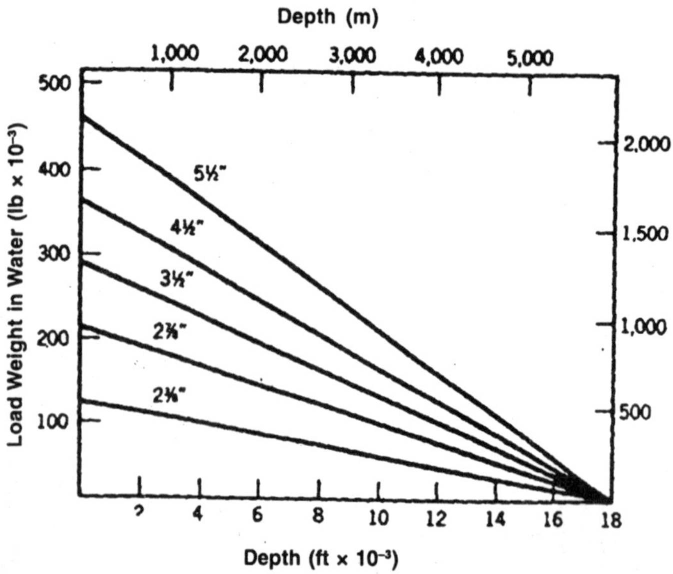


**Figure 13.2: Load-carrying Capability of API Drill Pipe**

of the ways, separated, and the lower half fabricated in a normal manner. The mating section of the upper half can be skidded sideways onto a parallel launching ways, then down to the outboard end, and the upper half fabricated concurrently. There are obviously several variations of the above, depending on yard layout.

Each half is now transported to a protected deepwater site, as was done with Hondo, and then launched. Note that the launching can be either end-O, or sideways, since the cross section is uniform. Mating afloat can follow the methods previously described for the Hondo platform, using stabbing cones and watertight access tubes, enabling welding in the dry. Temporary buoyancy tanks ensure that the tower floats horizontally on its upper legs. A spud-can section is constructed at the lower end of the tower.

After mating and the welding of the mated legs to form an integral structure, the auxiliary buoyancy towers at the lower end of the tower are ballasted to slightly negative buoyancy. Now the structure, having a slightly inclined attitude, is towed to its installation site.

Ballast added to the spud can section and lower jacket legs causes the structure to upend. Bending moments in the tower need to be carefully computed for this operation and may required progressive ballasting of mid-length compartments in the tower. The upper temporary buoyancy tanks are designed so that the structure will float vertically at the site. Further ballasting of the jacket legs and spud can cause the structure to touch bottom and then force the spud-can into the soil.

The pile anchors may then be drilled in, using a drill ship or semisubmersible, and then cemented. The first segment of the guy line will be attached to each anchor pile as it is set. Then the drill ship will layout the segment, lower the clump anchor, attach the second segment, and then lay it down. A pennant will be attached, with a marker buoy, so that the guy line can be retrieved when the tower is moved into position.

Once the tower has been set and the spud-can has achieved initial penetration, each guy line is fed in through a swiveling fairlead and run up to the deck and stopped off. A linear hoist is attached and initial tension taken up gradually around the series of guys. Then the upper temporary buoyancy tanks are ballasted to a slight negative buoyancy and removed.

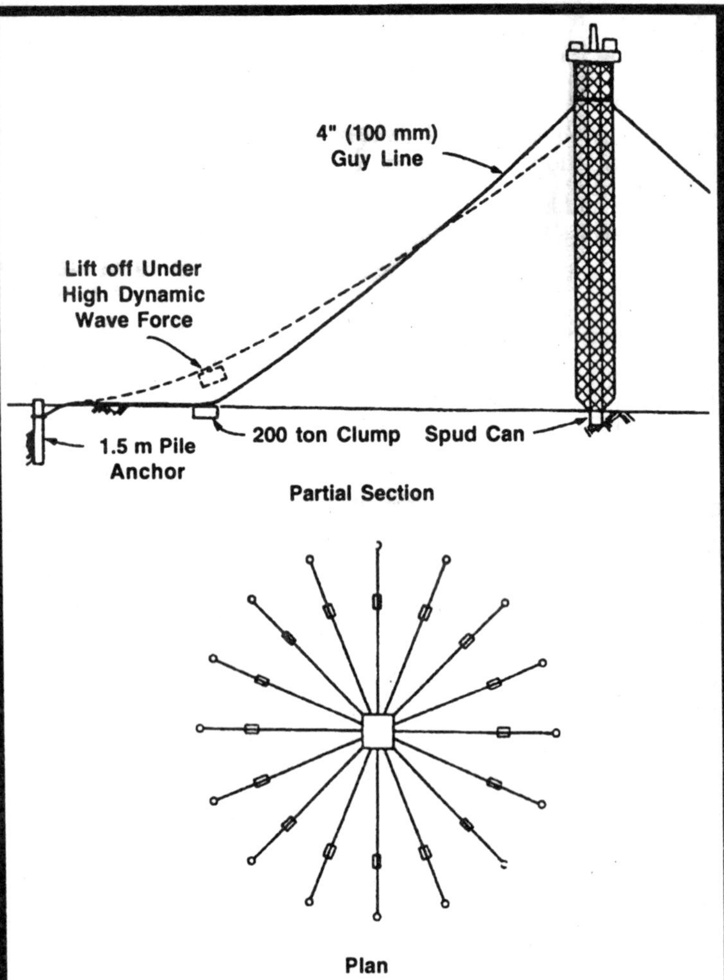


**Figure 12.3: The Guyed-tower Concept**

To help the penetration, drilling mud weighted with barites added to bentonite slurry is used to replace the water ballast in the spud-can. After penetration has been achieved, the heavyweight ballast can be replaced by seawater. Reducing the pressure in the spud-can will mobilize the hydrostatic pressure on its upper surface to further help penetration. Spud-can penetrations can be 15 m or more, depending on the soil stiffness. Suction anchors, developed since Lena, are even more effective in obtaining penetration of the spud-can.

Following final penetration, the tension in each leg is readjusted with the cable grip hoists. The deck structure and modules are then set by derrick barge. In other installations of guyed towers, a piled base structure was used, with the piles driven through sleeves in the guyed tower. This was the solution adopted for platform Lena, constructed in 1000 ft (300 m) of water, described in the following section.

Not only was platform Lena a unique structure once in place, but its installation embodied a number of new ideas which proved successful. The jacket was 330 m (1080 ft) long by 36 m (120 ft) square. Launch weight, including main piles and torsion piles, was 27,000 tons. The jacket was loaded out from the fabrication yard in conventional end-O fashion and then lowered onto transverse skids. Launching near the site was sideways, using four launch runners with guides plus rocker arms. Holdbacks were used to restrain the jacket while the barge was ballasted to heel 7° to starboard. Then the 80 mm frangible nuts of the holdbacks were severed by explosive detonation and hydraulic jacks activated to overcome starting friction. The jacket launch took only about 10s, one fourth the time normally required for a stern launch. The jacket had a maximum roll of 53°, the barge a maximum roll of 15°.

In all, 12 long buoyancy tanks, each 6 m in diameter by 36 m long, were built into the upper portion of the jacket. High-density iron ore slurry was placed into the base to assist upending. A derrick barge was used to control the upending, seating, piling, and guy line attachment. This derrick barge maintained position by means of four computer-controlled thrusters.

The lower sections of the main piles of the structure, 1350 mm in diameter, were carried out with the jacket. After upending of the jacket and its seating, they were extended and driven to 170 m penetration, making the completed piles almost 500 m in total length.

For the torsion piles, whose upper end terminated in the base, a system of latches and lugs was used to connect pile and hammer together for lowering as a single unit. One 100 mm-diameter multistrand wire rope was used to lower the combined unit, using a

600,000-lb capacity linear winch. Air, electric, and hydraulic lines were lowered with separate constant-tension winches and lines. Initially, hammer efficiency was reduced by the cushioning of compressed air below the hammer ram, but a change in the air exhaust system overcame this problem. After driving to full penetration, the latches and lugs were hydraulically released, enabling the hammer to be retrieved.

Earlier, the 20 guys and their anchors had been installed. Drilled in pile anchors were placed and grouted, each with the guy line preattached. These guy lines had articulated clump anchors attached. A barge laid out the guy with its clumps and then secured it to a buoy temporarily held in position by a small, taut gravity anchor. Once the tower was installed, connection was made with four lines from the tower, leading out through underwater fairleads. These were then pretensioned with linear jacks. Then the additional 16 guys were completed and tensions equalized. The guy lines were 135-mm-diameter wire lines, each 550 m in total length and sheathed in polyethylene. These guys had a breaking strength of 1525 tons each and were designed for 500 to 600 tons maximum load. The clump anchors were each 200 tons, consisting of articulated weights attached to the guy line segments.

## The Constructions on the Deep Seafloor

As structures are submerged, they can be positioned by dynamic thrusters, locked in by on-board computer through acoustic transponders to surface vessels and thence to satellites. Alternatively, they can use preset seafloor transponders to maintain relative position.

Submerged buoyant structures can be kept afloat at prescribed elevations off the seafloor by the use of weighted tethers. If they rise, they pick up more tether weight (for example, chain) and hence return to their original elevation. Objects lowered on rope or casing must consider the dynamic response of the lowering vessel as it responds to the waves, as well as the inertial effects of the object, with its added mass of water that must also be accelerated. To overcome this, giant heave compensators were devised for the, Glomar Explorer and were used to overcome roll, pitch, and heave effects.

Free-fall deployment of seafloor structures as well as anchors may also be "controlled" by installing multiple buoys. The structure can then be progressively ballasted to descend in steps. French engineers have developed "breather" buoys, which decrease in volume and hence buoyancy as they descend. These have been successfully used in laying a test section of pipeline in 2500-m water depth. The Harding GBT, a seafloor storage tank, was placed by the use of multiple buoys.

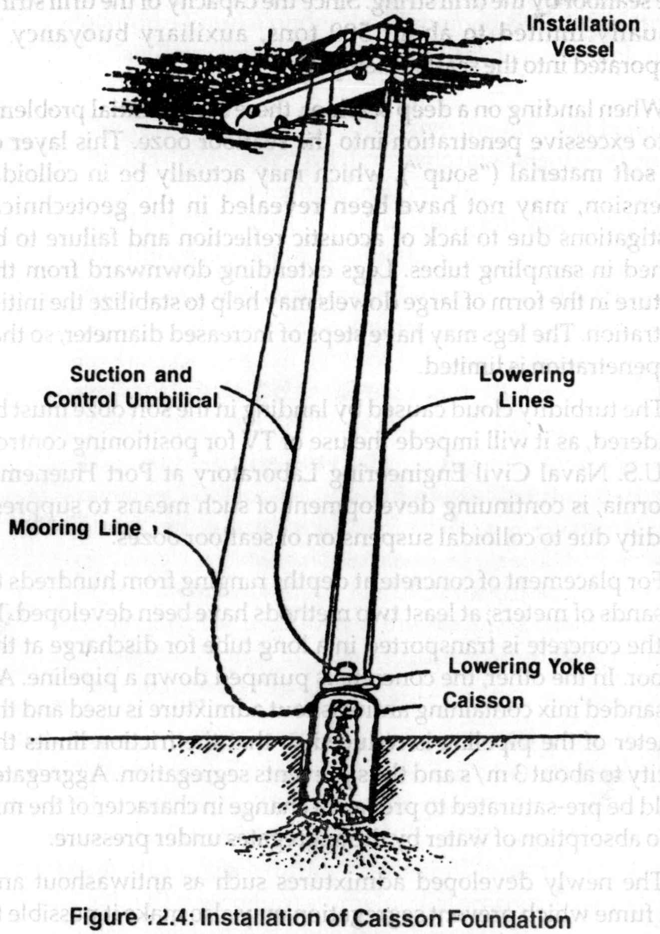


Figure : 2.4: Installation of Caisson Foundation (Suction Anchor) in Deep Sea

"Glide" can be used with submersibles and ROVs to control the rate of descent. "Pulling down" of a buoyant structure against a seafloor anchor is an effective system for decoupling the system from the surface wave effects once the structure is below the surface.

The latest generation of large offshore crane barges is equipped to lower seafloor templates to substantial depths approaching 1000 m. For these and greater depths, the seafloor template is submerged by crane barge to below the hull of a deep-sea drilling vessel, then slung in under the vessel's moon pool. It is then lowered to the seafloor by the drill string. Since the capacity of the drill string is usually limited to about 500 tons, auxiliary buoyancy is incorporated into the seafloor template.

When landing on a deep seafloor, there are potential problems due to excessive penetration into the seafloor ooze. This layer of very soft material ("soup"), which may actually be in colloidal suspension, may not have been revealed in the geotechnical investigations due to lack of acoustic reflection and failure to be retained in sampling tubes. Legs extending downward from the structure in the form of large dowels may help to stabilize the initial penetration. The legs may have steps of increased diameter, so that total penetration is limited.

The turbidity cloud caused by landing in the soft ooze must be considered, as it will impede the use of TV for positioning control. The U.S. Naval Civil Engineering Laboratory at Port Hueneme, California, is continuing development of such means to suppress turbidity due to colloidal suspension of seafloor oozes.

For placement of concrete at depths ranging from hundreds to thousands of meters, at least two methods have been developed. In one, the concrete is transported in a long tube for discharge at the seafloor. In the other, the concrete is pumped down a pipeline. An oversanded mix containing antiwashout admixture is used and the diameter of the pipeline is reduced so that the friction limits the velocity to about 3 m/s and thus prevents segregation. Aggregates should be pre-saturated to prevent a change in character of the mix due to absorption of water by the aggregates under pressure.

The newly developed admixtures such as antiwashout and silica fume which prevent segregation may also make it possible to place concrete underwater by use of closed buckets or other discrete

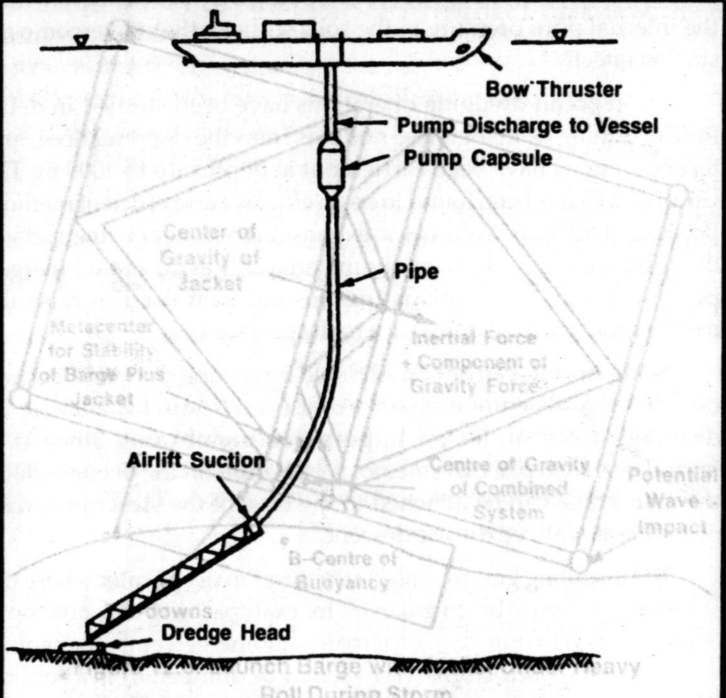


**Figure 12.5: Deep-ocean Dredging System**

devices. Cement slurries (grout) have long been placed by pumping at great depths by the oil-drilling industry, where they have been used to cement casing strings, and to plug wells. Concrete has been placed through tremie pipes to depths of 1000 m and more in mine shafts.

When concrete or grout is used in large volumes, the heat of hydration must be considered, and special cementing mixtures such as blast furnace slag cement or cement plus pozzolan must be employed to reduce the heat and prevent the consequent disruption of the concrete or grout. Grout is especially vulnerable due to the normally high cement content.

For breaking objects loose from the seafloor, waterflooding underneath is considered the most effective method. The pressure must be kept low enough to prevent piping to the sea. In soils of low permeability, many hours of such flooding may be required to raise the internal pore pressure in the soils sufficiently to overcome the suction effect.

Deep-ocean dredging operations have been studied in detail for the mining of manganese nodules from the deep seafloor, and test operations have been carried out at depths up to 4000 m. The use of airlifts has been found to be an effective and efficient method. Because of the large volumetric expansion of air near to the surface, the airlift is employed to raise the material only as far as a submerged pump capsule. Conventional pumps are then used to raise the nodules the additional distance to the surface vessel.

Seafloor soils can be consolidated by suction drainage, carried out after the structure is emplaced or even before installation, by drainage from under an impervious membrane. Since it is impracticable to place the membrane by itself on the deep seafloor, the membrane can be attached to the base of the structures, with pressures equalized during descent.

For installing gravity-based structures in deep water where the seafloor is known to be irregular–as, for example, with rock outcrops–one solution is to dump rock to create a submerged embankment on which to seat the structure. The rock may be dumped from a bottom-dump barge in one mass to minimize segregation during the descent. Alternatively, it may be placed through a flexible tremic tube; this latter provides better control. The rock should be pre-saturated to dispel all air. After dumping, further consolidation can be obtained by dynamic compaction (*i.e.*, the repeated dropping of a heavy ram or explosives). Screeding at depth is very difficult, so a preferred solution is to equip the gravity-based structure with long skirts and to equalize bearing by underbase grouting after landing.

The use of this flexible chute, suspended from a ship, provides control and prevents segregation. It may prove feasible to extend the depth range of this type of placement and thus provide a high degree of control.

To prevent excessive loss of grout through the rock embankment, a percentage of the rock fill should be of smaller size. Alternatively, smaller graded rock can be placed on top–some will be lost but other stones will fill the chinks between the large stones. The grout for underbase fill must have a thixotropic admixture to reduce its flowing tendency once the pressure drops.

Both of the above means were successfully used in shallower water under caissons for an offshore terminal in Queensland. These methods have been incorporated in a study for the deep piers (300 to 500 m) for the bridge across the Strait of Gibraltar.

For structures such as subsea templates seated on the deep seafloor, it may be necessary to transfer manipulators and service modules between them and a surface vessel. Pop-up buoys may be attached to such a structure, to be released on acoustic signal and thus provide a guide line for subsequently lowering or guiding a manipulator or structural element to an exact mating with the previously installed element.

Tensioned guide lines of this type were extensively employed in the 1960s and 1970s at water depths up to 500 m for re-entry of drilling strings into casing. They have now been largely replaced by acoustic and inertial guidance, as developed and used on the Glomar Challenger to reenter the casing at 6000 m.

As an illustration of how structures may be placed on the seafloor at great depth, the following example is given. A large gravity anchor block or underwater oil storage tank is to be placed at 6000 m depth. Procedures need to be developed. The structure is constructed as a steel-concrete sandwich (hybrid) design to have an impervious membrane on both faces and thus eliminate the problems of absorption of fluids under pressure. The structure is so configured that it can be completely filled with a low-density fluid such as a mixture of hexane and heptane and still remain afloat with minimal freeboard. In shallow water, the structure is submerged to sit on the seafloor by water ballast. A semi-submersible drilling vessel is floated over the structure and connected to casing running from the drilling rig. The structure is now snugged up under the semisubmersible and made fast for tow to the deep-water site.

On arrival, external ballast tanks are filled with high-density fluid (barite-weighted drilling mud) so that the structure is negatively buoyant. With calm seas, the structure is released to be suspended from the drill casing alone. The drill casing is in turn held by linear jacks or the derrick.

The structure is lowered by the drilling string. Pneumo-hydraulic motion compensators can be used to respond to the short-period relative motions. The most severe stresses and stress ranges will occur when the structure is passing through the shallow depths. These stresses generally reduce as the structure descends to deeper levels. As various depth horizons are reached, additional buoyancy fluid is added (under pressure) to offset the compression and reduction of volume in the fluid due to the increasing hydrostatic pressure and the increased weight of drill casing.

As the structure itself reduces in volume and the low-density fluid compresses further, some of the high density drilling mud (or iron sand) is discharged to keep the weight on the drill casing within allowable limits.

Upon touchdown, additional high-density drilling mud and/ or seawater is used to displace the low-density fluids in order to provide proper on-bottom stability. If this is a gravity structure such as a base for a TLP, for which weight is necessary, high-density drilling fluid or iron ore slurry should be used to fill the internal compartments.

Other lowering systems have been successfully employed in deep water. Both subsea drilling templates and undercarriages for ocean mining have been slung in under the "mother" vessel by the simple procedure of rigging wire rope lines, ballasting the template or carriage to slight negative buoyancy, and lowering it by means of a derrick barge or similar vessel. As it descends, it is drawn in under the mother vessel. The dynamic forces during lowering are absorbed by the wire rope lines. To increase their length and hence their stretch, they are made up in multiple parts. Neutrally buoyant fiber lines, such as nylon, have the added advantage of lower modulus of elasticity and hence greater ability to absorb dynamic loads. In any event, a complete dynamic analysis needs to be made of the coupled systems to ensure against harmonic resonant response at all stages of descent. A similar procedure has been proposed for assembly of the 30-m-diameter OTEC cold-water pipe.

These systems, along with floating production systems and tension leg platforms, are emerging as the trend in development of oil and gas in the deep sea. Methods for mooring of floating systems such as SPARS or special-purpose vessels, are being developed along several lines: the use of preset spread moorings in which the deadweight is partially offset by buoyancy, either empty drill casing or syntactic foam blocks, dynamic positioning using diesel fuel initially and process gas as it becomes available, and the use of

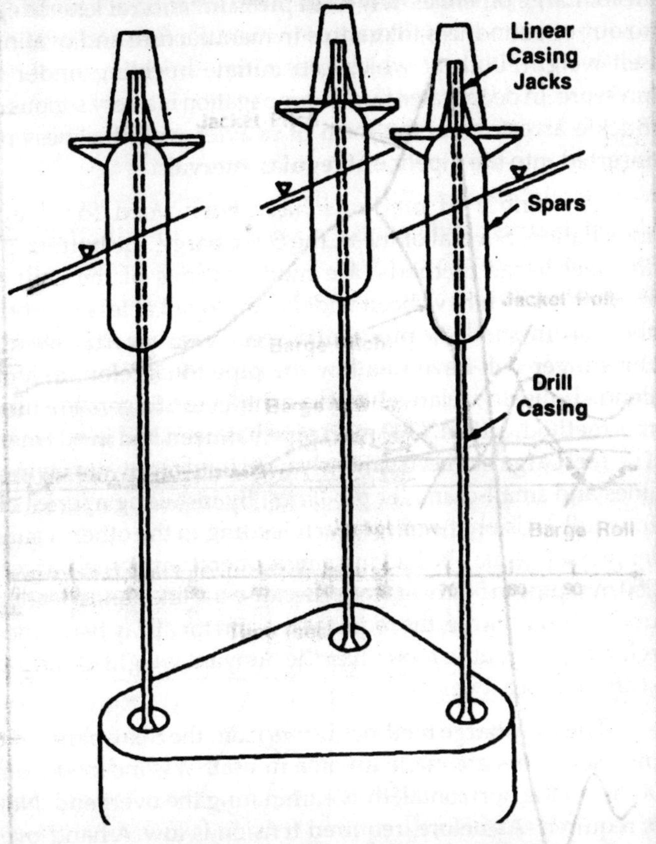


**Figure 12.6: Lowering a Heavy Mass in the Deep Sea**

high-strength drill casing. Recent developments and successful installations such as those described in this chapter clearly show the capability of advanced technology for operating in the deep ocean.

## The Deep-water Pipelines

Deep-water pipelines and flow lines have been successfully laid in water depths as great as 1700 m (Mensa project). Even greater depths are in the planning stage. Below 1000 m, a typical pipeline has to be internally pressurized to balance the external hydrostatic head. Large pipelines may need pressurization at lesser depths due to out-of-roundness tolerances in manufacture and ovaling due to self-weight, both of which can initiate buckling under external pressure. In deep water, buckle propagation is a very serious concern. Buckle arrestors, in the form of heavier wall thickness pipe, are inserted into the pipeline at regular intervals.

A number of methods have been used for deep-water installation. Several of these, the S-lay barge, the bottom-pull, and the reel barge methods, are modifications of the conventional methods which have been widely employed at lesser depths. With the S-lay method, the pipe will descend nearly vertically and hence the stinger will have to allow the pipe to develop an almost 90° bend. Typically, relatively short cantilever stingers are used. With this method, a 12 in. (300 mm) pipe has been laid in 600 m of water. The reel barge has been similarly used for deep-water laying of flow lines and small-diameter pipelines. By installing a series of two to four full reels on the barge, each feeding to the other, a substantial length of line may be laid in a single operation. A reel barge has laid 250 mm pipe in 500 m of water. Because the line hangs nearly vertical under the lay barge, the required tension for all lay barge methods is relatively low, little more than the buoyant weight of the pipe from seafloor to sea level.

The J-lay barge method differs from the S-lay barge in that the pipe segments are made up on a ramp that is inclined from 60° to 90° from the horizontal, thus eliminating the overbend. No stinger is required. As before, required tension is low. A hand-over-hand tensioner has been mounted in the inclined ramp. The joint has to be made and completed at one station, located just above the deck. Hence, advanced rapid means of welding are employed. The most efficient and rapid laying is attained by racking up pre-assembled triple and quadruple lengths of pipe.

The J-lay method is a highly specialized, highly developed method for laying long major pipelines in deep water and rough seas. Since there is no longer a significant horizontal line tension, dynamic thrusters can be used to position and move the barge, eliminating the continual anchor handling required for third-generation S-lay barges. The J-lay system is planned for use on the Ursa project in 1300 m of water and designs are now available for J-lay pipe from 4 in. (100 mm) diameter to 14 in. (350 mm), in 1600 m water depth. Modular systems are being constructed to permit the installation of J-lay ramps on existing offshore pipe-laying barges without extensive modification to the barge.

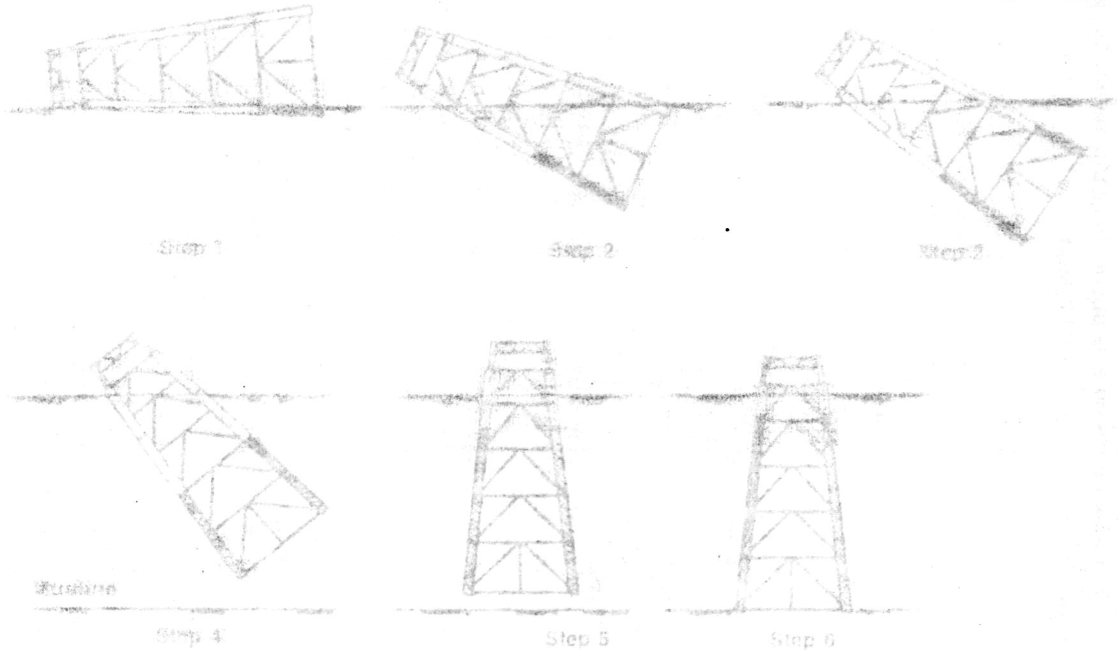


Figure 12.10: Installation of Self-floating Jacket

# Index

N

**T**